T0275881

THE CRISIS IN PHYSICS

CHRISTOPHER CAUDWELL

Edited with an Introduction by
PROFESSOR H. LEVY

VERSO

This edition published by Verso 2017
First published by John Lane The Bodley Head 1939

© Christopher Caudwell
Introduction © H. Levy

1 3 5 7 9 10 8 6 4 2

Verso
UK: 6 Meard Street, London W1F 0EG
US: 388 Atlantic Ave, Brooklyn, NY 11217
versobooks.com

Verso is the imprint of New Left Books

ISBN-13: 978-1-78663-460-3
ISBN-13: 978-1-78663-461-0 (UK EBK)
ISBN-13: 978-1-78663-462-7 (US EBK)

British Library Cataloguing in Publication Data
A catalogue record for this book is available from the British Library

Library of Congress Cataloging-in-Publication Data
A catalog record for this book is available from the Library of Congress

Printed in the United States

CONTENTS

iii

CONTENTS

INTRODUCTION

WHEN the present international nightmare has passed, and the day of reckoning has come, Man may be able to sit down calmly to count the cost of his ignorance, his stupidity, and his social mal-organization, in terms of human suffering and social sacrifice. Not least among these will be the destruction of men of great promise— in cases, amounting to genius—on the Spanish anti-fascist front. It was no sordid motive that drove the International Brigade to take up arms. It was alien to their nature. They were individuals with a heightened social awareness. If ever men consciously sacrificed themselves that others might live, these men did. Because they were capable of this great renunciation, and because their action was dictated by intellectual conviction, they represented the flower of our youth, alive, alert, understanding, sensitive.

Christopher Caudwell was killed in Spain. A young man still in his twenties, without great educational advantages or technical training, he has nevertheless left behind him a mass of written material of such deep understanding as to mark him out, had he but lived to enjoy the society for which he died, as one of our most gifted men. It is inevitable that there should be many such buried on Spanish soil, for it was precisely men of his quality who early realized the meaning of that desperate struggle.

What had the crisis in physics to do with Christopher Caudwell? What had the crisis in physics to do with the writer of *Studies in a Dying Culture* and of *Illusion and Reality*? In what way were these linked in his mind? In what way were they related in Nature? How could the problems of technical and philosophical significance with which modern physics was wrestling—Relativity, the Quantum Theory—stir one whose mind appeared to move in a totally different plane? In what possible sense could he have anything valuable to contribute to the solution of these complex problems?

Christopher Caudwell had burrowed far beneath the surface of events. To him it was no mere accident that the struggles in the West, that had broadened so far as to involve and unsettle every European state, should also have penetrated to such depths as to obtrude themselves into every aspect of social life. When our western economy is rending itself in twain, western society challenged by what it has itself created, when the great knowledge and power which it has built up is being thrown into the deepening struggle, it is obvious also that the minds of men will scan with critical eyes as never before the logical and emotional basis of their activities and beliefs. Each in his own way, each even within his own special domain, will seek the understanding that will lead him to a way out of the greater crisis.

A period of social instability, whatever the underlying causes, must sooner or later call a halt to those ventures that can proceed only when society is moving on an even keel. A long-continued slump must arouse poverty and disquiet, not only among unemployed craftsmen and among those who might have catered for their needs, but also among that great body of executive officers of capitalism whose administrative powers are being brought to nought. The tacit assumptions, the judgments and valuations they have accepted in the past, are finally dragged to the light of day in the search for the necessary understanding. Traditions that have flowered in periods of quiescent development are cast aside, and new outlooks engendered. The minds of men are on the move. As the logic of capitalist economy works its way towards war, as the best intellectual and physical energies of communities are bent towards the creation of more and more efficient weapons of destruction, so the ethical assumptions that have been distilled from the experience of the past become frustrated and negated, and socially valuable activity is damped down. The present western economy, at the highest point of its development, can persist only in decay, only by an increasing renunciation of the best of what it has produced. It involves a denial of the freedom to expand. It cannot solve the problems it arouses. It enters a period of decline, of contraction, of frustration, of restriction.

Those who, like Christopher Caudwell, have been sensitive to all that is best in Man's creations in the past, conscious of the power Man has acquired over Nature, alive to the inner meaning of Man's achievements, revolt against this Dying Culture, against the forces that stifle progress. They strip the Illusion from the Reality. Not alone the cries of those who are destroyed to-day but the call of unborn generations sound in their ears, appealing to redeem the world while there is yet time. Every aspect of this vast struggle throws up its problem for solution, its cultural side, its aesthetic side, its scientific side. Men

INTRODUCTION

like Christopher Caudwell force their way to where the
fire burns fiercest, and for the greatest of all causes give
up their lives. The struggle has sharpened their understanding, for a sharpened understanding was necessary to
ensue it; but the cost is heavy.

Before Man could act in his capacity as a physicist he
had first to be a social being; there can be no science without a social background. But every society moves forward
on certain tacit assumptions that, for a time at least, remain
imperative, unchallenged. They are deeply embedded in
the accepted outlook of the men and women who carry
through its social activity. Every social problem is
unconsciously approached from this standpoint. It is the
origin of analysis. It is this outlook that in its time is
applied towards the resolution of the scientific problems
that are encountered in the effort to carry forward the
work of that society. It embodies the relatively obvious,
the things that cannot be questioned. From it, there are
drawn images and concepts which, when pieced together
as a pattern, provide the conscious theoretical ground-
work of each period. On this basis, therefore, any scientific theory is necessarily the specialized development of
a general social view, even although those who initiate
the theory may be profoundly unaware of the connection.

From such a socio-philosophical background the
scientist therefore tends, in the first instance, to arrange
his theories in terms of the categories he unconsciously
applies in social life. This is part of Christopher Caudwell's thesis, and if it is true it is of profound significance.
In what way then does this understructure of scientific
theory begin to manifest change? The pursuit of science,
in the first place guided and interpreted in terms of these

images and concepts, leads to a growing body of knowledge that finally must outstrip the tacit assumptions that are basic in them, no matter how deep-seated. Driven on by the necessities of practice, scientific men struggle to recreate new images—at times even doing violence to so-called common sense, in the effort to resolve their difficulties—to reconcile theory and practice. This is the first mode of change.

There is yet another way in which this movement occurs. These images and concepts of fundamentally social origin represent an aspect of the prevailing ideology. Its form depends on the socio-economic structure. When, therefore, economic instability sets in, the ideology of that social phase moves from unconscious acceptance to conscious criticism. Just as soon as the categories of social life begin themselves to shift, as in the present, so also, therefore, will a movement of a similar nature be reflected within the inner structure of theoretical science. A crisis in Society will reflect itself indeed in a crisis in ideology, and in a series of crises in diverse branches of science and art. All theories become the subject of fundamental criticism. It is in this way that the linkage between science and society reflects itself in the formulation of theory. A deep-seated social crisis involves in its turn a corresponding unsettlement in every developed branch of science.

This is a thesis which in a vague and general way has at last come to be widely recognized. The mutual conditioning of science and society has become itself an accepted category, but this has not happened until the nature of the relationship is rapidly changing. Nevertheless the mere fact of its acceptance is already evidence that the transition in outlook among scientists has begun. In a period of prosperity and economic development science, it is seen, also expands. In a period of contraction it is restricted and frustrated. Step by step it marches in general well-being or in decline with society itself.

Such a formulation, however, has already become a mere truism to-day, so rapidly have the minds of men been affected and to that extent become sharpened, by their economic uncertainty. With Christopher Caudwell the analysis penetrates deeper. What, he asks, are the tacit assumptions on which bourgeois society in the past has developed, and on the basis of which it has built up its traditions? To him the answer is patent. In effect, he asserts, the week is split in twain; on the Sabbath the minds of the people are concerned with Man's inhumanity to Man, during week-days the energies of Man are devoted to economic exploitation of his fellow creatures. The former represents the humane, the subjective, the emotional—the latter the machine; the former the mental and spiritual, the latter the material. Thus the economy of bourgeois society is riven by an internal contradiction. Its economic basis, which studies and treats of Man and machine in identical ways, is essentially, therefore, mechanical. The scientist, who is concerned with such matters, becomes therefore in scientific practice a mechanist. The mental qualities in Man find no place as objects in his analysis. The philosopher, on the other hand, has no interest in matter; to him the mental and emotional characteristics of man are all that is of importance. He is a subjectivist, an idealist.

And so society in the capitalist era contains two conflicting ideologies—mechanical materialism on the scientific side and idealism on the philosophical side; and these have arisen because the working practice of bourgeois society demands that they shall be split in twain. It is the subject-object relationship that underlies the present epoch. It provides the fundamental categories of modern society.

A period of crisis emerges. The social traditions of the scientist, his ethical beliefs that have in uncritical fashion been built up around the philosophically idealist outlook which he has accepted—simply because they are socially accepted—become frustrated. The ruthlessness of the machine does violence to what has become in that setting his finer self; and yet the perpetuation of his scientific work, on the same mechanistic basis as before, cannot do other than accentuate the very factors that are doing violence to his feelings. In being driven on to the study of Man as machine, Man as a cog in the machinery of bourgeois production, he is faced with an internal contradiction between his social theory and his scientific practice. That is the first level at which the crisis in science shows itself, and no solution can be forthcoming for him until a unity is achieved between these two opposites—objective mechanism and subjective idealism—a unity that will bring out their mutual conditioning, their mutual development, and that will expose the patterns of social development that must necessarily emerge from their interplay. He has to be emancipated from the limitations of a science that can regard Man only as a machine and from a philosophy that can acclaim him only as an idea. This in itself is drastic enough, for it drives him to no less than the study and the practice of social change. It is for that reason that already scientific men number among their ranks many of the most politically conscious members of the community.

The crisis in science does not cease at this stage; it must proceed ever deeper. Even the world of material change is not mechanistic. To adopt a machine-like theory in the analysis of matter is to pass the problems of Nature through a mesh and to concentrate one's attention only on those that are left behind, while the others escape. This evasion of Nature, however, is only in theory. In practice the problems left unsolved impinge on Man's experience and force him to their study and analysis. In the end Man must dominate Nature. In such a situation his mechanistic theories are found wanting. Without a clear appreciation of why he has come to accept a mechanistic standpoint in the past, and therefore without an anticipation of what its limitations may turn out to be, he will presently find himself faced with yet another contradiction—that between theory and experimental practice. That indeed is the situation in the world of physics to-day; and that crisis which is emerging in the realms of Relativity and in Quantum Theory, at the macroscopic and the microscopic levels, is therefore fundamentally a partial aspect of the whole crisis in bourgeois economy.

To appreciate these facts and to adjust his mind to the new modes of thinking required to resolve them, demanded in Christopher Caudwell a combined social and scientific understanding that would be rare in a scientist of mature experience; to find them in this young man is almost phenomenal.

In one sense the present work was never completed. It consists of twelve chapters; the first six of these were left in comparatively finished form, the remaining six, eminently readable as they are, were mainly in rough draft and without chapter headings. I have not ventured to modify this in any respect, preferring rather that the book should be produced as it was laid down by Christopher Caudwell when he left for Spain. In the body of the manuscript there have appeared, here and there, short phrases and cryptic notes, clearly intended as reminders to himself, of the points he proposed making. In general, when these have occurred near the beginning of a chapter, I have placed them in italics at the head; otherwise I have left them in italics in the body of the chapter, just where they appeared in the rough manuscript. Beyond this I have made no alterations whatsoever.

H. LEVY

CHAPTER ONE

FROM NEWTON TO EINSTEIN

1. THE NEW SCHOOL IN PHYSICS

THE crisis in physics, which a few years ago was the secret of physicists, has now become generally shared with the public. Even the man in the street is aware that all is not well with physics; and that in many cases the cracks which are rapidly developing in the structure have been stopped up by mystical notions new to science. It is proclaimed by distinguished physicists that 'determinism' or 'causality' has been expelled from physics; that the Universe is the creation of a mathematician; and that its real nature is unknowable. Jeans, Schrödinger, Heisenberg, Dirac and Eddington are prominently associated with these ideas; all are distinguished physicists. They are opposed by Planck and Einstein, whose prestige is the chief weapon in their defence of the older positions. For their defence is a kind of stone walling; they are unable to lead any counter-attack on the enemy positions. Planck's justification of 'causality' is that it is the scientist's faith, his anchor, the unprovable fundamental of science. Einstein's tactics are even simpler; he 'cannot understand' what the younger men mean.

Evidently the new school do not need to trouble about dislodging their antagonists from such ineffective philosophical positions and, with the support of the bishops

3

and the spiritualists, they advance to occupy the new territory they have marked out. Of course it is impossible to ignore the opposition of Planck and Einstein. Einstein is the father of relativity physics and Planck of the originator of quantum physics. Both were 'revolutionary' in their day. Even Planck's faith and Einstein's incomprehension therefore have pulling power over the undecided. But the younger men include Heisenberg, Schrödinger and Dirac whose technical achievements are of a similarly 'revolutionary' character. There is no doubt that the new school is winning mass support in its struggle for a more mysterious Universe.

The cause of the crisis in bourgeois physics is sometimes held to be the contradiction between macroscopic or relativity physics on the one hand, and quantum or atomic physics on the other. The concepts with which each domain works are irreconcilable. But it would be wrong to suppose that this contradiction is the real cause of the present crisis in physics. The crisis is too general for that. This particular contradiction is only one of the forms in which the crisis comes to light.

2. NEWTON'S UNIVERSE

There has in fact been a contradiction between two domains of physics ever since the days of Huyghens. Newton's system of Nature, which included the corpuscular theory of light, formed a consistent scheme of the Universe, apparently free from contradictions, built up on an atomistic basis. All particles behaved according to a simple law of motion which uniquely determined the life-line of each particle. The system was of such a character that an 'initial push-off' and an initial fabrication of the atoms out of nothing was necessary. These initial acts were creative acts of God. God thus appears in the Universe as force and substance alienated from Himself. But once created, these two categories are subject to law, the laws of the conservation of matter and energy. Given its initial push-off and creation, the atomistic universe is self-running.

Newton however does not regard it in this light, for his conception of substance is such, as we shall see later, that the maintenance of these laws in fact requires the continual intervention of God.

Thus such a Universe does not exclude the possibility of divine interference with its own laws, but it is always a disruption of very simple laws, and hence is bound increasingly to appear an unaesthetic act.

In the medieval and Aristotelian schemes of the Universe, motion requires the constant expenditure of force, apart altogether from laws governing the action of forces. Hence the Universe needs the continual inflow of Divinity, as a Prime Mover, to keep it going. Evidently therefore Newton's atomistic scheme gives a basis for deleting God from the Universe as a causal influence once it is treated. The laws of God then become qualities of matter. As compared with Aristotle's, Newton's laws of motion desacralize physics; and they culminate in Laplace's divine calculator, who, knowing the speed and location of every particle in the Universe at a given time, can predict the whole future course of events throughout infinite Time. Nature becomes a machine, but of course one can still ask with Paley: 'Who made the machine?'

Newtonian physics excludes God from Nature, but

not from Reality, because it makes Nature only a part of Reality as a result of its particulate conception of Matter.

3. THE WAVE THEORY OF LIGHT

The experimental disproof of the corpuscular theory of light shattered this Universe in the eighteenth century and Laplace's divine calculator had in fact already been proved an impossibility before he emerged from the brain of the French mathematician. It was proved that light rays did not have the character of corpuscles but of waves.

Now everyone had seen waves, and therefore there seemed nothing startling in this conception. But waves as witnessed are waves in something: they are a certain type of movement of water particles. But in the succeeding years, light waves, although they continued to behave like waves in water, proved to be waves of nothing. This raised problems of a critical kind, but the deepness of the contradiction and the gravity of the crisis were only gradually realized.

It is true that this nothing was given a name: the ether. Ether it was explained was not matter; its properties were *sui generis*. Unfortunately all these *sui generis* properties proved to be negative. Ether offered no resistance to matter. Ether had no chemical properties. Ether was frictionless, weightless, invisible, and unaffected by the passage of matter through it.

Its final and utter negativity was revealed by the Michelson-Morley experiments. Since the one certain *sui generis* property of the ether was that light waved in it, then at least a property peculiar to light waves in motion could be recorded of it: the speed of this motion as compared to the earth's. An ingenious apparatus was constructed based on this argument: The earth moves through the ether; light waves are waves in the ether; hence if the movement of light waves relative to the earth across a given distance is measured first across the earth's path and then with the earth's path there will be a discrepancy. This discrepancy will show the earth's real speed through the ether.

In fact the result was null. There was no discrepancy. The logical assumption was that the ether moved at the same speed as the earth. But could the earth possibly drag all the ether of infinite space with it? This was contradicted by observations of the stars; and the phenomenon of 'aberration.' These observations, and also experiments with 'ether-whirling' machines, excluded the only logical deduction from the experiment; that bodies dragged along with them the ether in their vicinity.

4. THE BEGINNING OF THE CRISIS

Hence physicists were faced with the proof that light waves were waves of nothing—evidently an unacceptable statement, for it is meaningless. The only escape from this was a theory that circumstances always mysteriously changed to prevent their observing the earth's motion through the ether. This alternative was adopted under the name of the Fitzgerald contraction. It was assumed that matter moving through the ether contracted along the line of advance so as exactly to conceal the very discrepancy of measurement, which would reveal the speed of the earth.

7

This conception was not so fantastic as it sounds, for meanwhile matter had revealed electro-magnetic qualities, and electric and magnetic fields had been proved to obey a set of equations, developed by Clerk-Maxwell, which also controlled the emission of light. Light waves were special forms of electro-magnetic waves. Analysis of the electro-magnetic equations showed that they might be interpreted to mean that matter would contract to the required extent as a result of its motion through the ether. The Fitzgerald contraction was widely accepted as a fact of Nature, and the solution of the crisis.

Meanwhile the nature of ether remained unknown; its specification included factors that insured its unknowability. Science found on its hands that metaphysically unmanageable entity, the unknowable.

For in fact the unknowable cannot exist; even to say that it is unknowable is to say we know something about it; and when further we say it is unknowable for certain reasons (as we must if unknowable is to be more than a mere word) we specify certain of its qualities, although in an inverted way.

If this position was to be taken seriously, either the ether was completely unknowable and therefore did not exist, being merely the nominative of 'to undulate' or else relative motion through the ether was unknowable, in which case this too did not exist. In either case this unknowability defined certain definite characters of the knowable entities, light and motion. 'Omnia determinatio est negatio.'

5. RELATIVE AND ABSOLUTE MOTION

This revealed a contradiction which was already extant in the Newtonian scheme, whereas the other contradiction had emerged as a result of the discovery empirically of what were held to be undulatory characteristics in light. All the Newtonian particles were in motion, and for example each particle's velocity gave its kinetic energy, if squared and multiplied by its mass. Its energy and mass therefore seemed real self-subsistent entities. But no particle can move in relation to itself, only in relation to something else. Thus a car moving along a road at 30 m.p.h. encounters another at 30 m.p.h. moving in the opposite direction. Relative to each other they are moving at 60 m.p.h. However, we say each is 'really' moving at 30 m.p.h. because that is their speed in relation to the earth of which the road is a part. But the earth is turning on its own axis and circling the sun; and therefore that car which moves with the earth's rotation and orbital motion is, in relation to the sun, travelling some thousands of m.p.h. faster than the other car. Indeed in relation to the sun, a more important body than the earth, the car is not moving forward at all, but hurtling backwards. Yet the sun is not fixed, but itself moves in relation to the stars, and these themselves move in relation to each other. Hence unless some body at absolute rest can be found, it is impossible to find the true speed of any particle, and hence its energy, and hence its inertia and hence its mass. These can only be found relatively, and in any case, even if such a body at absolute rest does exist, the mass, energy and inertia are still relative and not self-subsistent. Only the resting body

9

could be regarded as self-subsistent. Newton realized these difficulties in a general way and only talked of bodies absolutely at rest with the proviso, 'if any such exist.'

Now if the speed of the earth through the ether could have been determined, then the ether could have been assumed to be at absolute rest, and this would provide a cosmic framework for detecting the absolute of 'true' motion of all particles. But we have seen that motion produced the Fitzgerald contraction, exactly concealing the velocity.

However this Fitzgerald contraction itself conceals a contradiction. The length of a body through the ether contracts as a result of its motion. But this in itself implies an 'absolute' length, which is the length of a body at rest in the ether. But since it is impossible to establish the rest or motion of a body in the ether, absolute length is as unknowable as absolute motion. Since the Fitzgerald contraction is unknowable, it cannot be held really to exist. It is merely another negative determination of moving bodies.

Motion *includes* time: a certain space is traversed in a certain time. But in concrete reality time is not built up into motion. Motion is 'broken down' into time. The movement of a body is, in a clock, analysed into movement in space and duration of time. Hence if absolute or time motion and length (or space) are both unknowable, then this is equally so of absolute time, for the motion of bodies will be broken down into different components of space and time by different observers.

The ultimate conclusion of a chain of reasoning which

we have only briefly indicated here was that the absolute dimensions, time, and velocity, energy and mass of any particle were unknowable. They did not exist in themselves, or in relation to a unique framework, but were properties of relative frameworks.

6. THE SPECIAL PRINCIPLE OF RELATIVITY

Einstein recognized that these unknowabilities were in fact important principles of knowledge about nature, and he formulated them as the special Principle of Relativity.

This states that absolute length, mass, energy, space, time, and motion do not exist. But before this Principle could be formulated as a scientific principle and not a metaphysical doctrine, it was necessary to establish the relativity of these qualities in a practical way. Although the Principle of Relativity has an epistemological content, it is not a principle of epistemology, but of science. It describes the limits of our knowledge about reality in such a way that these limits become real descriptions of the nature of matter in relation to us. This was only possible because the previous experiments which had established these limits, had furnished a fund of real knowledge about Nature. This fund could not be used by the existing theory of Nature. On the contrary, these practical results contradicted this theory, which therefore had to be recast in a form fuller of practice.

As long ago as Lucretius philosophers have advanced theories as to the relativity of motion and the secondary and defendent character of abstract Time.[1] But all

---

[1] 'Time also exists, not by itself but simply from the things that happen.' Lucretius.

11

such theories were purely metaphysical and could be countered by opposing theories of equal logical worth. It was because the Special Principle of Relativity co-ordinated and gave a meaning to a mass of empirical observations, that it was of importance to physics, and made deeper man's understanding of the Universe.

7. UNITY AND ATOMICITY

Yet this at once brought to light a still older contradiction, which had also been immanent in the Newtonian Scheme. The Newtonian 'bodies' were self-contained units which had each been created with an initial mass and an initial packet of kinetic energy in the form of mass multiplied by the square of the speed which enabled them to lead a wholly independent life in the shape of a right line. Unless they collided physically with another particle, the existence of each was self-contained and unchequered. In such a Universe, unless a collision took place, nothing 'happened' and even such a happening merely meant that the two particles continued on right lines at different angles and speeds. Happenings in such a Universe are therefore completely accidental in this sense, that they represent the intersections of two chains of events (the 'life-lines' of the particles) which are self-contained and self-subsistent. They are also completely predetermined in that, given the relative positions and velocities of the particles at any time in their history, it would have been possible to predict their collision with certainty.

Such a Universe is of course completely pluralistic. It has no organic unity. The history of the other particles has no effect on the history of one. From the point of view of the particles all happenings are complete accidents. From the point of view of observers of the particles, all happenings are completely predetermined necessities.

Such an ideal Universe is however only partly the Universe of Newton, which already contains another unifying principle, as 'mysterious' and 'transcendent' as God, contradicting the atomism of the Universe. This mysterious principle is rendered necessary by observation. In fact none of the particles travel on right lines but all are more or less curved by the effects of the other particles. This curvature is therefore of gravity, an intangible entity whose real nature is unknowable—it can only be expressed in terms of its 'effect' on the paths of the particles, which it causes to curve towards each other in different degrees, the shape of the curve depending on the mass-velocity of the particles concerned.

Since this force affects all particles, it is as resolutely monistic as the other conception is pluralistic. In this sense no particle's path is self-contained for to specify it with perfect accuracy, the mass and location of every other particle in the Universe must be known. Thus no happening—no collision of particles—is entirely accidental, for in the life of every particle the lives of all other particles have been bound up from the start, and no collision is a collision of two absolutely independent chains of events. For the same reason no event is completely predetermined, for to estimate it, all precedent events must be taken into consideration by the calculator, whose own consideration therefore becomes an element

13

in the problem, provoking a new situation, making it as insoluble as if a man were to try to climb to a height great enough to look down on himself.[1]

This principle appears to be something apart from the qualities of matter, which are all self-subsistent in the individual particle. In the Newtonian scheme each particle is a complete individualist, unrolling from its past history, its complete future fate, even though that fate may be continually interrupted by accidents (collisions). But the force of gravity is a kind of omnipresent Power, apparently non-material, since it acts across a distance. Indeed, it is evident that to Newton all action of this kind is closely associated with the idea of God. Our subsequent examination of seventeenth century metaphysics will show that this whole atomic Universe was built on the hypothesis of God. Hence the force of gravity already appears as the result of a metaphysic which divides the Universe into matter and non-matter. This had important consequences for the subsequent development of physics. The Newtonian combination of monism and atomicity had this logical defect, that it stated certain laws of motion, which determined uniquely the lifelines of all particles. Then to these laws it added the proviso, in the form of the Law of Gravity, that these Laws could never be obeyed, for another force applying to particles between themselves would always modify these laws relating to particles in-themselves.

In the Newtonian Scheme, the quality which carries on the particle in its independent life-line is 'inertia.' That

[1] This is Planck's argument in favour of free will and I have quoted it as an example of the deepest understanding of necessity to which mechanical materialism can attain.

14

quality which everywhere alters or distorts this life-line from the path it should follow as an independent unit is 'mass.'

'Inertia' is therefore the quality determined by the laws which govern the independent motion of individual particles; mass is the quality determined by the laws which govern the mutual attraction of particles. These laws are expressed by their effects on each other. The laws of motion produce a distortion of gravitational behaviour, as in centrifugal action. The law of gravity produces a distortion of inertial behaviour, as in gravitational force. And yet, by an apparently amazing coincidence, inertia is always equal to mass.

Although this statement endured for over two centuries, evidently there is something gravely suspicious about its formulation. The very facts that inertia and mass are equal and that one set of laws is expressed in terms of deviation from the other set, and vice versa, points overwhelmingly towards a synthesis of these laws into a common set. Yet one—the set of laws regarding motion —is based on the conception of the Universe as composed of independent particles of matter. The other— the gravitational law—gives us a Universe which is an all-containing force of Unification, where the shudder of a leaf on earth is reflected in a corresponding alteration of gravitational forces on Sirius. Evidently then the required synthesis must

(a) Reduce mass and potential energy and inertia and kinetic energy to a common basis.

(b) Express the laws of motion and of gravity as derivatives from one fundamental law.

(c) Reconcile the atomicity of matter particles[1] with the monism of gravitational attraction.

8. THE GENERAL PRINCIPLE OF RELATIVITY

Such a synthesis remains a mere theory as long as it is based only on logical considerations. But meanwhile a number of physical discoveries had intervened to give the contradiction an observational basis. These facts had already led to the Special Theory of Relativity. The Special Theory denied the existence of absolute distance —yet the Newtonian force of gravity is a product of distance and mass. Hence if a universal force is the product of distance and mass, and distance is relative to the observer, mass must be relative to the observer too. Again, since the force of gravity appears as acceleration, or change of motion in objects, and according to the Special Principle of Relativity all uniform motion is relative, how can change of motion be absolute?

The Special Principle of Relativity therefore, when once established, made necessary the General Principle. Just as the Special Principle states that absolute uniform motion and length do not exist, so the General Principle of Relativity states that absolute change of motion and mass do not exist. Moreover it satisfies the problems (a) (b) and (c) tabulated above as follows:

(a) Mass, momentum, kinetic and potential, energy and inertia, are all different forms of the 'inertial' quality of matter. They can all be expressed in the frame of a common geometry, which is not Euclidean geometry but 'real' geometry. Or, put in another way,

[1] 'Mass-points' in the technical vocabulary of physics.

Euclidean geometry is only real in certain special circumstances.

(b) This 'real' geometry synthesizes the Newtonian law of gravity and the Newtonian laws of motion in one basic law: 'The directed radius is constant in empty space.'

(c) The behaviour of the particle is determined by the geometry of the rest of the Universe. In other words Einstein's world is monistic, and eliminates the pluralism of the Newtonian system.

The geometry of this Universe is the geometry of a *continuum*. It has no absolute Space and absolute Time, but these are welded into one block geometrically and each observer will divide the block differently into space and time; no division will be absolute or unique.

Relativity physics does not make all qualities relative —it sets them in a new absolute framework. Interval— in which both space and time figure—takes the place of distance as an absolute separation between events. The velocity of light is an absolute velocity, whatever the observer. The amount of matter in the Universe is absolutely constant and the conservation of momentum still holds absolutely as a law of Nature. And the absolute framework in which these new qualities are set, is the continuum of space-time, specified by four dimensions.

If this is the real world then it is plain that the logical incompatibilities and distortions of the Newtonian world are due to the fact that the continuum has been split up into an absolute Space-in-itself in which the individual particles move, their movements being accompanied by the uniform flow of an absolutely Universal Time.

However, we have no reason to accept one theory

17   c

because it is more synthetic than another. The important point about the Principle of Relativity is that if it is correct, objects would observably behave differently in certain circumstances from what they would if the Newtonian theories are correct. These differences have been observed and support the theories of relativity physics, which are therefore accepted by physics, to-day. This is proof that Einstein's theories are truer than Newton's: it is not a proof that Einstein's theories are absolutely true—a belief that would obviate the need for further study in this domain of physics.

Although the dimensions which Newton had supposed to be absolute were shown by Einstein to be relative, this does not therefore mean that Einstein believes all dimensions of the Universe to be relative. On the contrary, his whole life's work has been devoted to eliminating relative qualities from physics in order to reach at last a firm absolute foundation. Each revelation of a relativity in dimensions was regarded as a crisis which could only be solved by 'restoring normality' on a new plane—in other words, by again putting physics upon an absolute basis. Relativity in dimensions or qualities is regarded as a kind of unreality and illusory subjectivity about them, which is opposed to the absolute character of objective reality. Absoluteness and relativity are regarded as mutually exclusive qualities.

Now this is a metaphysical assumption. It is an assumption common to Newton and Einstein. The difference between them is the fund of new information about the observable behaviour of objective reality which forces Einstein to damn far more qualities of matter with the label 'relative,' than was found necessary by Newton.

It also forces him to look far more deeply into the structure of the Universe in order to find absolute qualities.

The crisis of physics is not therefore the result of any contradictions in relativity physics, or its supersession of Newtonian physics. Relativity physics is all of a piece with Newtonian physics. At every stage contradictions already latent have become open as a result of extended observation of Nature; and at every emergence they have been resolved by means of a new theory which lifted physics to a higher plane. The contradiction between the Galilean laws of motion and the Keplerian laws of planetary motion, led to the Newtonian equation of mass to inertia and the formulation of the Universal law of gravity.

At a later stage the wave theory of light emerged as a contradiction of the particulate theory of matter, and attempts to resolve it not only gave birth to field physics and the electro-magnetic equations of Clerk-Maxwell; but also pointed the way to the modern developments of atomic physics.

To-day, however, the integrations are becoming increasingly unstable.

The solution of the contradictions within mechanics by the relativity theory, and the solution of the contradictions within 'wave' physics[1] by the electro-magnetic equations of Clerk-Maxwell, and the solution of the contradictions within atomic physics by the quantum theory, has only led to greater contradictions between these three domains of physics. Conditions call impera-

[1] i.e. Field physics, covering electromagnetic phenomena, including light.

19

tively for a synthesis of the laws governing the three domains, but each new discovery makes this less likely, and the conflict more acute. It is this which has given rise to the present crisis in physics and made it wholly different from previous crises, which merely paved the way for an immediate synthesis. Here however far more drastic revision is necessary. It is significant that in discussing the consequences of these contradictions, scientists find themselves forced to discuss concepts such as free-will and the nature of knowledge which had hitherto been excluded from science as philosophical questions. The scientist in other words is compelled to overhaul his philosophy, which hitherto had been an uncritical and inevitable way of looking at things rather than a conscience metaphysics. It was none the less metaphysics. Indeed, because of this unconsciousness, it was all the more metaphysical.

This overhauling of their philosophy by scientists has been singularly unsuccessful. The very fact that it has been undertaken, however, is a sign that this crisis is different from the previous crises of physics. It is a revolutionary crisis.

A revolutionary crisis occurs when the contradictions discovered in practice, cannot be met by a rearrangement of content within the categories of the domain of ideology concerned. The categories of this domain are in turn dependent on those of other domains of ideology and a revolutionary crisis is the signal that no real solution is possible, unless the most basic and fundamental of categories, those common to all domains of ideology, are more or less rapidly transformed. Hence the crisis 'overflows' from physics into other fields.

Einstein and Planck are the last physicists who accept the old metaphysics of science uncompromisingly, and who therefore attempt to site their empirical discoveries in an ordered world-view. They are the last physicists sharing the philosophy of Newton and Galileo, although of course it is a philosophy transformed by all that has taken place in the meanwhile, transformed, but not revolutionized. Einstein and Planck are the last of the solid 'Old Guard' of Newtonian physics.

CHAPTER TWO

THE WORLD AS MACHINE

1. REVOLUTION AND MYSTICISM

THE integrations achieved by Einstein and Planck in their respective domains, gave rise to a contradiction between the domains which burst asunder the much-patched fabric of physics. This is realized by the 'younger men'—Jeans, Eddington, Heisenberg, and Schrödinger. But it would be wrong to regard this new school as revolutionary in a real sense—as men who can renew the fabric of physics. For it is the essence of a revolution that such a transformation can only take place as part of the transformation of the fabric in all fields of ideology, and this in turn is part of a still deeper transformation.

The physicists we have mentioned show no realization of the fact that there is a causal connection between the crisis in physics, and the crisis in biology, psychology, economics, morals, politics and life as a whole. Where they see a connection it is only the connection of a general 'disease' or 'questioning.' Thus to adopt a genuine revolutionary standpoint in physical theory involves the adoption of a genuinely revolutionary attitude in real life. This is not what any of the 'new men' are guilty of, although in all branches of science still newer men are emerging who show traces of just this

22

solid revolutionary position.[1] The older antagonists of Einstein and Planck are aware of the untenability of the metaphysics of current science, but their attitude is purely destructive: 'down with all metaphysics!' They regard this as a progressive step. But of course it is in fact impossible to have a theory without a philosophy: the philosophy is implied in the theory. It is impossible even to have a practice without a theory: one is implied in the other. Hence the slogan, 'Down with metaphysics' which also takes the form 'Down with concepts,' or 'images,' or 'theories,' leads on the one hand to a narrowing and specialization in the field of physics, so as to keep it apart from life as a whole—a so-called empirical and positivist attitude towards science—and on the other hand the exclusion of science from their own general world-view which thus becomes mystical and idealistic. In other words there is a cleavage of theory and practice—practice becomes specialized, restricted and empirical, and theory becomes abstract, unco-ordinated and diffuse. Hence in spite of an increase in technical competence in the particular field they have made their own, there is a reactionary trend in their general world-view, which regresses to forms left behind by science.

Once begun this cleavage accelerates. As practice becomes more specialized, and bare of theory, it becomes more difficult to integrate the different specializations in one consistent world-view; and ideology as a whole becomes more anarchic and confused.

Because of this Einstein stands out as a larger figure than his successors, because of his possession of a clear

[1] For example, in biology, J. B. S. Haldane, J. Needham and L. Hogben.

and all-embracing philosophy which was able to contain a wide domain of physics. His philosophy, however, was not adequate to contain and synthesize the whole complexity of modern physics, whose anarchy it has indeed helped to produce. The pending revolution in physics is therefore the incursion of a wider philosophy able to contain the various specialized and contradictory domains, and resolve them into a larger synthesis.

2. THE METAPHYSICS OF SCIENCE

It would be absurd to suppose that this philosophy can come into physics except from outside. The present metaphysics of physics were not generated by physics, but physics were generated by it, not in a self-contained way, but by interaction with reality.

The present metaphysics of physics—its philosophy— did not descend into physics from the air. Before man could function as a physicist he had to live as a man, and not as an abstract man, but as a real man in a certain society. If we take modern physics, as beginning with Galileo and Bacon, then the physicist was a man who had to live as a member of bourgeois society before he could function as a physicist. To do so was to have a whole superstructure of theory, conscious and unconscious, generated by participating in all the myriad functions of a real man in real society of that kind. To be a physicist was to apply this most general theory to a particular domain of quality in reality, that of physics.

What one would find was determined on the one hand by the nature of reality, and on the other hand by the theory brought to bear on reality. The operation

24

would be a selective one, and the selection would be mutually determined by one's theory and reality's nature. The impact would be mediated by instruments, and these in turn would depend on the technical level of the society in question, and the resources it could spare for research.

It may be argued that this does not allow for the 'genius.' But in fact the theory of the physicist is not stamped on him but is the resultant of a tension between his innate qualities and his experience of society. None the less his qualities can only be realized through the categories of society and thus emerge with the grain of the epoch, however carved. The greater the genius, the more profoundly he will be penetrated with the qualities of his experience. In science this means, the greater the genius, the more penetrative of Nature the categories of society will become in his hands. The theory of a man is his world-view, and ultimately informs and guides his every action—is in fact inseparable from it. It may not however be realized consciously as a world-view. Any new theory, such as a scientific hypothesis, because it is an extension of his world-view, necessarily is arranged within its categories, even if the arrangement brings about some transformation of them. Hence the genius does not escape from the categories of his age, any more than man escapes from time and space, but the measure of his genius consists in the degree to which he fills these categories with content—a degree which may even result in their explosion. This explosion is, however, in turn dependent on a certain ripeness in the categories.

Physics separates itself out from the web of thought and action, but remains in organic connection with its matrix. Bourgeois physics is completely contained within the categories of a bourgeois world-view and when it escapes from them even Einstein 'cannot understand' it. But it can only so escape in a crisis when the web itself is breaking up.

3. THE ECONOMIC CRISIS

It is no accident therefore that the crisis in physics occurs at the same time as an unprecedent economic crisis, which has become world-wide. The crisis is not peculiar to physics, it penetrates all ideology. In its most general form it is the growth of anarchy by and through integration; it is the explosive struggle of content with form. In the words of Planck:

'We are living in a very singular moment of history. It is a moment of crisis, in the literal sense of that word. In every branch of our spiritual and material civilization we seem to have arrived at a critical turning point. This spirit shows itself not only in the actual state of public affairs but also in the general attitude towards fundamental values in personal and social life.

'. . . Formerly it was only religion, especially in its doctrinal and moral systems, that was the object of sceptical attack. Then the iconoclast began to shatter the ideals and principles that had hitherto been accepted in the province of art. Now he has invaded the temple of science. There is scarcely a scientific axiom that is not nowadays denied by somebody. And at the same time almost any nonsensical theory that may be put forward in the name of science would be almost sure to find believers and disciples somewhere or other.'[1]

[1] M. Planck, *Where is Science Going?* 1933.

These words reveal a general feeling of collapse of the old order, together with a complete helplessness and lack of understanding as to its cause, which is characteristic of certain elements of society in a revolutionary crisis. Everything is confused, culture is tumbling about his ears: that is all Planck knows.

The symptoms are precisely the same in all spheres of ideology. There is an increasing specialization and technical efficiency inside the different domains of ideology, but this leads to an increasing anarchy and contradiction between the domains. It is not merely that biology separates from psychology, but psychology itself splits up into mutually exclusive disciplines. Hence it is no longer possible to have a synthetic world-view, a living theory in touch at all fronts with practice. The theory is forcibly torn apart. In such circumstances there are three alternative attitudes open to conservatism: (a) A mystical positivistic attitude to all spheres of ideology outside one's little garden (Eddington); (b) A violent reduction of all other forms of thought to the highly limited categories of one's small domain (Freud); (c) An eclectic mish-mash of all the various specializations with no attempt to resolve their contradictions. This leads to a world-view that negates and frustrates itself (Wells). Obviously any of these alternatives merely intensifies the crisis.

But in fact this ideological anarchy is only a reflection of the economic anarchy which is the cause of the general crisis. When I say 'reflection' I mean that the same general development has taken place in the sphere of social relations as in ideological categories, because the latter are merely subtilizations, qualitatively different, of the former.

27

It is the characteristic of bourgeois economy that its social relations contain a contradiction which brings about its development and also its decline. This contradiction is the contradiction between socially-organized labour, on the one hand, and individual appropriation of the products of that labour, on the other hand. In its early stages this is the only means by which a raising of the productive forces beyond the stage of handicraft can be accomplished.

A time is reached however when increasing organization of labour within the factory, with its tremendously increased productivity, leads to violent conflict between the individually owned factories. This is the imperialistic stage of capitalism: the era of increasing competition between the trusts and monopolies and the nations which are their organized expression. It becomes plain that the social relations are holding back the productive forces and this is apparent in 'over-production,' mass unemployment, slumps, and wars. Humanity is driven forward to revolutionize the productive relations of capitalism, to set free the crippled productive forces. Capitalism turns into its opposite, communism.

The whole superstructure of ideology, which is in active relation with its base, is thus more or less rapidly transformed, and the new categories generated lead to a synthetic world-view on a higher plane. Of course it is not suggested that physical theory is a mirror-reflex of social relations. It gives information about non-social reality. But it gives such information *to* society. The knowledge is conscious knowledge. It has therefore to be cast into the categories of society.

These categories are not like the Kantian categories,

eternal and given in the nature of mind, a set of tools which work up into a cognizable shape the unknowable thing-in-itself. Man interpenetrates actively with Nature. The depth of his interpenetration is due to the fact that he works in association. The laws of association, in the most general sense, are therefore the dynamic field along which individual men actively struggle with the object. This struggle is not merely physical—practical—it is also theoretical, a relation of cognition. Only in abstraction can the two be separated. Hence the social relations are reflected in all the products of society (including the ideology of physics) as categories.

Physics is knowledge about reality, but it is abstract, generalized knowledge. The abstractions or generalizations are the reflections of the social relations by and through which the reality was made into conscious knowledge. Some of these categories are general to all society; but they appear in a special form in different societies, and evidently in the case of the crisis in bourgeois physics, it is the specifically bourgeois categories that are of vital interest, because of the way in which new knowledge, new practical content, is rending them asunder, and is itself crippled by the old form.

4. MATTER AND MECHANISM

The unconscious philosophy of the contemporary physicist is mechanism. When the bourgeois considers matter as the object of cognition, he is unable to conceive of it except under the categories of mechanism. The categories of mechanism are: atomism, 'strict' causality, absolute time and space. Outside these categories, the object is unknowable to the bourgeois philosopher: hence if like Kant he regards these categories as creations of the human mind, matter-in-itself becomes unknowable.

Matter is a name for the category of objective reality. The field of physics is objective reality in its most generalized form. Historically, as with Aristotle, the field of physics included all 'Nature'—i.e. all matter. But gradually certain qualities of matter were excluded from physics, e.g. those of biology and chemistry—and it became bourgeois physics.

The philosophy of physics is the philosophy of all bourgeois in relation to matter. It is mechanical materialism. The philosophy of all bourgeois philosophers in relation to matter is the same; but for various historical reasons bourgeois philosophers ceased to be interested in matter, and developed another part of bourgeois philosophy, that concerned with the mind or subjective reality. This they regarded as 'real' philosophy, distinct from physics. Hence what is called to-day, philosophy, is only a section of the true bourgeois philosophy or world-view.

It is equally true that the mechanism of physicists is only a part of their philosophy for they also accept the standard bourgeois world-view in regard to mind, that of idealism. But just as the 'philosopher' is not interested in matter, the physicist is not interested in mind.

In the main, therefore, physicists and philosophers share a general bourgeois world-view in which the physicists concentrate on developing one department, that of matter, or objective reality, and the philosophers that of mind, or subjective reality. The bourgeois philosophy of subjective reality cannot escape from the standpoint of idealism or conceptualism. Hence bour-

30

geois ideology, in all fields, reveals this cleavage between subjective reality and objective reality as a struggle or contradiction between mechanism and idealism, matter and mind, causality and free will. This is the notorious subject-object relation, the most famous problem of bourgeois thought.

5. THE WORLD OF BOURGEOIS SOCIETY

Is it possible that this cleavage has any connection with the basic contradiction of capitalist economy, which secures its development and decline? Could it be that in the sphere of ideology a contradiction, reflecting the cleavage of the foundation, has first of all unfolded all the complexities of bourgeois ideology, and is now causing them to disintegrate in anarchy? In fact there is apparent a close connection between the two.

In feudal society man is subordinate to man. Serfs and land, the medieval means of production, are owned by the ruling class, which also exerts coercive rights (the feudal dues, monopolies and privileges) over the bourgeoisie.

The bourgeoisie secures the abrogation of all 'rights' of man over man, and substitutes for it merely a right to own and dispose freely of things, including one's own labour power. This involves the shattering of all feudal restrictions and the creation of the 'free market' for commodities. Formerly only a small surplus of the goods a man produced came on the market; the majority were for his own consumption. Now not only do all his products become commodities, but many things hitherto thought inalienable—his faith, his loyalty, and his truth —require a cash value too.

31

By this means social productivity is raised to a new high level. The social division of labour is carried to an unprecedented degree; and it involves of course a corresponding social organization of labour. The two are not exclusive, but are opposites which produce each other. Specialization involves integration. Where each commodity is produced from start to finish by one man in his home, no complex social organization of labour is necessary. When into the making of each product a complex chain of separate specialized processes has entered, including the making of machinery, transport, and central control, then a corresponding social organization of labour is necessary: the organization of the factory and the town.

The organization of the factory is conscious—planned and controlled from start to finish. But the sum of factories which constitutes society is not so controlled, but their working is controlled by 'laws' of supply and demand—that is, the free market. The free market was the condition for the establishment of bourgeois society.

What are the laws of the free market, which hold together the producers of a society based on commodity production?

'Every society based on commodity production has the peculiarity that in it the producers have lost control of their social relationships. Each produces for himself, with the means of production which happen to be at his disposal and in order to satisfy his individual needs through the medium of exchange. No one knows how much of the article he produces is coming on to the market, and how much demand there is for it; no one

32

knows whether his individual product will meet a real need, whether he will cover his costs, or even be able to sell at all. Anarchy reigns in social production. But commodity production, like all other forms of production, has its own laws, which are inherent in and inseparable from it; and these laws assert themselves in spite of anarchy, in and through anarchy. These laws are manifested in the sole form of social relationship which continues to exist, in exchange, and enforce themselves on the individual producers as compulsory laws of competition. At first, therefore, they are unknown even to these producers, and have to be discovered by them gradually, only through long experience. They assert themselves therefore, apart from the producers and against the producers, as the natural laws of their form of production, working blindly. The product dominates the producers.'[1]

Evidently, therefore, there is a contradiction between the organized centres of production and the disorganization of social labour as a whole due to the interposition of the 'free' market. But this 'disorganization' is not a mere lack of organization, it is the specific form of society in a bourgeois economy. What stands between the organized centres of production are the rights of individual owners, whose life and freedom depends on their rights to extract a share of the value of the goods produced by the means of production owned by them. This share is not extracted immediately, when the goods are produced, but only when this value has been realized in the free market. Hence both the individual ownership and

[1] Engel's *Anti-Dühring*.

the free market are necessities for the existence of the bourgeoisie and of bourgeois economy, and their categories permeate all bourgeois society. The blindness in society as a disorganized whole is the inverse of the special status of the bourgeoisie.

The means of production must be worked by men, and since coercive ownership of men by men is abolished with feudal society, the bourgeois has no direct coercive ownership over men. But in fact men who own nothing can only live by bringing their labour power to the free market, because the means of production, without which man in capitalist society cannot realize his labour power in products, are owned by the bourgeoisie. Hence the capitalist's coercive ownership of things in fact veils a coercive ownership of men; but in appearance bourgeois society is one in which man has not 'rights' over man, only over things—i.e. over Nature. His right over Nature which is also his freedom, is in bourgeois theory realized passively, as a simple property right. But man's right over Nature is in fact realized by the improved production technique, an increasing division of labour, which is also the source of the real conscious organization of society. This division of labour is not based on ideal categories or religious hierarchies, but on the laws of Nature as these are discovered; the stuff having to be treated in such and such a way, and to go through this and that process, to realize the end desired.

Hence certain complexes are formed in society which constitute its organization—machines, nests of machines and the men arranged round them in certain active relations. The exact structure of these organizations of men depends on the necessity of Nature.

Of course these complexes are called for by certain wants but these wants are uncontrollable in bourgeois society—they merely emerge from the blind market. Once emerged, their satisfaction is wholly dependent on the intrinsic properties of Nature—it is a technico-productive problem.

In fact man's desires are also subject to necessity. They change with history, with the change of methods of production and corresponding alterations in the superstructure of society. Yesterday a Roman glutton; to-day an Egyptian hermit. But all this causation of desire in society is hidden by the basic form of modern society, in which desires emerge from the blind market.

Now society is the struggle of Man and Nature. In the more refined ideology of society this basic struggle appears as the basic problem of the subject-object relation. Man is the subject; Nature is the object. Therefore in bourgeois society, the object appears solely as 'things' over which man has rights, and whose laws or 'necessity' he discovers in order to satisfy his desires. These desires appear arbitrarily proposing an end for Nature to satisfy, and by exploring the necessity of Nature, they are satisfied.

Notice that these desires for products appear spontaneously, and the products, having been formed, disappear. The desires come out of the blind market, and vanish into it. And yet the market veils the desires of Man, his whole active relation to Nature, as a conative creature, and veils also the satisfaction of those desires, which take place behind the same screen. Hence the object is split off from the subject, and Nature appears as something wholly independent from Man. Nature is the object of determinism; she is the domain of Necessity.

Man desires certain things of her; and by making use of the known laws of her determinism, these desires are gratified. Man is subject to spontaneity.

Nature is always known as a passive object—as something not subject to man's activity nor the antagonist of his striving, but as something self-contained and shut in by its necessities. Hence Man's whole relation to Nature bears the stamp of the property relation, in which his right over it is the reward of his consciousness or cleverness—never of his activity.

The growth of the productive forces under bourgeois economy is an indication of the success of this conception of Nature. Nature's necessity becomes increasingly known. But this conception of Nature known by Man, is Nature known as a machine.

6. THE WORLD AS MACHINE

According to the bourgeois, the machine is a piece of Nature obeying deterministic laws so designed as to satisfy his wants and create use-value. It is as it were a self-contained piece of Nature which fulfils a 'plan.' The plan is his desire. This plan is to him something spontaneous and external to Nature.

Therefore the categories of objective reality in bourgeois philosophy are categories of the machine as it realizes itself in bourgeois society. The world is a machine —as machines seem to the bourgeoisie. It is the bourgeois himself who invented the term 'mechanism' and thus gave away the economic determination of his categories of objective reality. He had come to know Nature via

36

the machine, hence the laws of Nature came to him to seem identical with the laws of the machine. He explored nature by means of the technical development embodied in the machine—whether the machine of the factory or the laboratory.

The point is, the machine is not just a piece of Nature. It is a piece of Nature associated in bourgeois society with human beings who work the machine. It is the kernel of a social complex which gives it its shape and significance. When the bourgeois sees Nature as a machine, therefore, he sees Nature stamped with all the special and transitory social categories which that complex bears when viewed from his special standpoint in society. Nature looks a little queer to the bourgeois because he has a peculiar standpoint in society from which the machine too looks a little queer, and yet it is only through the machine that Nature enters into the consciousness of society.

We are criticizing the bourgeois philosophy because its view of Nature is 'mechanism.' That does not mean we believe that Nature's laws are different in kind from those of a machine. In fact this would be an absurd suggestion; since a machine is constructed out of bits of nature, according to natural laws, the laws of the machine cannot be wholly different in kind from those of Nature. Therefore what we are ultimately criticizing in the bourgeois philosophy of Nature is not the application of categories drawn from the machine to Nature, but the error in the bourgeois view of the machine. The bourgeois conception of the machine is at fault. That is why we say the categories of bourgeois scientific knowledge are economic categories, although it is knowledge about wild Nature.

37

The bourgeois theory of the machine is based on the part he plays in relation to the machine in concrete living. We have already seen that his role, and the relations of bourgeois economy, are such that man's desires emerge from the night of the market and are realized through the machine as products, which vanish again into the night. This is commodity-production, the basis of bourgeois economy in factory art and philosophy.

7. THE MACHINE AS SLAVE

Hence it seems as if Man's desires are altogether independent of the machine and that Man as it were stands outside Nature, like a visitor to the aquarium outside a tank. He observes the movements of the fishes. But the glass screen which cuts him off is a one-way screen. He can make his desires realized by the movement of the fishes, but the fishes' movements do not affect his desires. His relation to Nature is god-like. She serves his end like a slave. Nature, the machine, takes the place of the slave, servilely realizing the will of the master, with this difference, that the master must know her inner necessity.

This qualification, it must be admitted, marks an advance on earlier civilization. The slave obeyed the master's will, and because they were both men, it was not necessary for the master to know the slave's inner necessity, his capabilities and law, for these were crudely realized already by an inner and instinctive sympathy. Like master, like man.

But even so, this godlike detachment of man from machine is an illusion. For this godlike survey of the

38

machine overlooks the man who works the machine. Yet a machine without a man to work it is meaningless, since the machine is stamped through and through with operability. And as man's control over Nature by means of machines increases, so does the organization of society; and the lives of most of its units more and more dominatingly reflect the interpenetration of society by the machine. More and more men are organized by the necessities of struggling mechanically with Nature. Only the owning class escapes from the bonds of this organization, and is so much the more ignorant in blood and bone of the nature of reality. Hence mechanism is an illusion peculiar to the ruling class. The men who work the machine realize that so far from its expressing a one-way relation of Man to Nature, it expresses an each-way determining relation of which they get the full brunt. The laws of machines, of production, determine their whole lives. These laws determine the complexes of organization which, taken as a whole, make up bourgeois society.

The machine as a visible entity is only the kernel of the factory-complex with all its outgrowths, but the organization it crystallizes, the natural necessary laws of production based on division of labour shape the whole social hierarchy of the factory. It roughens their hands; determines their leisure; bows their backs; limits their horizon. Their very life depends on its activity; their relations to their fellows, to society; their freedom and their marriages and their friendships, are determined by the complex at whose heart is the machine.

Thus so far from the proletariat—the major part of society—standing in god-like isolation from the machine and Nature, their existence is determined by it; they are arranged about it like iron filings along the 'lines of force' round a magnet. For they *work* the machine; they form one producing complex with it. They cannot regard Nature as a passive shut-in object of contemplation.

8. THE BOURGEOIS AS OWNER

But the bourgeois owns the machine. Ownership is a one-way relation in so far as it enters the consciousness of Man. The other relation, the ownership of a man by property, which stamps a man with its characteristics, is unconscious. Formally to own a thing means that it has every obligation to me, and I have none to it. Any obligations to it imposed on me by law, custom or morality are regarded as so many limitations of ownership. Absolute ownership would exclude them.

Hence the bourgeois has in relation to the machine and the complex it produces that god-like isolation which is postulated in his theory of the machine. The wheel of the machine revolves, carrying round with it the proletariat who serve it and urge it forward; but it grinds out for its owner 'automatically' the freedom to gratify his desires in the market. He stands by; gives it perhaps an overseeing eye; or goes away for a holiday, knowing that it will continue to turn. Eventually he absents himself altogether; a chain of formalities, stocks and shares, veil even the turning machine from his vision.

But to suppose that any one-way relation between things is possible in reality is a fallacy. All relations are mutually determining; it is unthinkable that cause and effect should not mutually determine each other.

40

I own the stick; I wield it; but there is a reaction on my hand. The stick is my property; I am equally a stick owner.

But one determinism may be conscious; the other unconscious. The ownership is conscious; the being an owner is unconscious.

Thus the bourgeois is precisely aware of the way in which Nature, in the form of the machine, fulfils the desires emerging from the market and so gives him the means to fulfil his own desires. But he is unaware of the way in which the machine determines the movement of the desires of men.

9. CLASS AND WORLD-VIEW

The machine is a piece of humanized nature. It is composed of particles arranged according to a plan, the plan of a human desire. But the society which uses the machine, is a naturalized society. It is composed of men organized according to a plan, the plan of production. The organization of capitalist society, its factories, transport, and all the social grouping produced by this, is imposed upon it by 'the division of labour,' that is, by the necessities of Nature when these are operated upon by man to fulfil his desires.

But the bourgeois does not consciously plan the organization of society. It emerges blindly; it crystallizes out from the centres of production as the crystal of a super-saturated solution form on wires dipped into it. The warp and woof of the organized society of capitalist economy is spun blindly—by the growth of the machine under the blind laws of the market. Hence capitalist society presents the unique picture of disorganization amid organization.

Hence even the bourgeois is subject to the machine. But he is subject to it in a different way from the proletariat. The proletariat sees its subjection directly. The bourgeois owns the machine, and therefore the worker must sell his labour power to get into the factory and produce his means of subsistence. Once in the factory, his existence, his work, his co-operation with his fellows, is determined by the evident needs of machine operation.

Hence he has two relations to it not indissolubly connected; (*a*) a precarious and coerced relation to the machine due to the capitalist's ownership of it, (*b*) and the natural relation to it springing from the nature of machines. The first is obviously arbitrary and a matter of special social privileges. It is distinguishable from the latter, which is given in the very nature of life, since a machine is a machine, and must be greased, repaired and fed. Both these ways in which the machine determines the life of the proletariat directly are overt and conscious. One included all the weaknesses of capitalist society; its anarchy, slumps and mass unemployment; the other all its advances—its increased productivity and complex web of economy.

The bourgeois however is subject to the machine in an unconscious and veiled way. His suppressed determination by the machine is forced into the blindness of the market. The way in which production by its immanent laws determines the whole organization and movement of society is in bourgeois economy veiled by the market, and these laws do not appear as laws of the machine in relation to the bourgeois but as a totally new set of laws, the laws of supply and demand, and the laws of capitalist competition. But these laws are in fact lawless, since the capitalist never knows how much to produce or what will be the fate of the product. He never knows which way the market will turn, which is to say that he does not know the laws of the market. It is anarchic.

Yet in bourgeois economy the market is the only way in which human desires can appear as active forces realizing themselves, and dictating the machine process. Hence human desire appears to the bourgeois as 'spontaneous,' that is, anarchic and undetermined or certainly not as determined by the machine, whose laws (as he thinks) he precisely knows. Hence the subject and object have become completely separated. On one side is the man, desirous, active and spontaneous, that is, subject to no law, emerging freely, and wholly undetermined by the machine. On the other side the object, Nature as known by Man, the machine, contemplated in splendid isolation, whose mere contemplation secures the subservience of Nature to him. This cleavage does not seem to the bourgeois odd; any other arrangement of it seems unconceivable. He cannot imagine himself being free if the spontaneity of human desires, or the independent mechanism of nature, were in any way infringed. And in this respect he is right, for these particular forms express the sole conditions which can secure the existence of the bourgeois class as a privileged class. If therefore they were abolished, the bourgeois class would cease to exist and there would be no more bourgeois freedom.

10. THE SHATTERING OF A WORLD-VIEW

The fact that this schematization of the subject-object relation contains a contradiction becomes increasingly evident. It is just the effort to resolve this contradiction which secures the development of bourgeois ideology in philosophy and physics, just as the same contradiction in the form of individual ownership and appropriation (spontaneous irresponsible desire) and organized social labour (the objective laws of nature) brings about the development of bourgeois economy. Only finally the contradiction shatters its own categories and emerges in a synthesis: in economy, communism, in ideology, dialectical materialism.

The bourgeois feels the determining influence of the machine via the market in an increasingly coercive way. It appears as slumps, a general economic crisis, unemployment, currency chaos, over-production, a necessity driving him to war. And he gradually comes to feel that the 'machine has got out of control' and expresses this in vague desires for a close season for invention, limitation of plant, and rationalization. However he is not conscious of the precise way in which the machine determines his life as for example the proletariat is, because it determines it via the 'free' market and the 'free' market consists in its unconsciousness. His refusal to become conscious is not however merely wooden obstinacy, and pigheadedness. For consciousness is not a mere contemplation, it is the result of an active process, which in this case would imply active control of the market. But the market is merely the net result of the actions of individual producers for it. Hence active control of the market would involve active control of the whole process of production and therefore the extinction of his right of individual ownership, which is the condition of his existence as a bourgeois. Thus it is not mere obstinacy, but a life-and-death matter. For society to become conscious of the determining relations of the machine upon itself in their fullness, the bourgeoisie must cease to exist as a class. No wonder therefore that the subject-object relation is as insoluble by bourgeois philosophers as it was by philosophers of earlier class societies.

Therefore the categories of mechanism, which are the only categories the bourgeois philosopher is able to apply to nature, are categories of the machine in the special way it functions in bourgeois economy. He is unable to achieve any other categories since even those of teleology, which is put forward as an alternative, are, as we shall presently show, precisely the same as those of mechanism.

11. THE SHARING OF MATTER

Physics is concerned with *objective* reality, with nature, with matters behaving as a machine is supposed to behave. In physics nature is studied in a glass tank—the physicist merely wanders on and surveys the scene. Thus nature in the struggle of man with nature appears as the object *in contemplation,* the object as it is in itself, measured in terms of its own necessity. Such an object is quantitative, bare of quality, and hence the Nature of bourgeois physics is bare of quality.

This stripping was a gradual process. Matter to Galileo and Bacon is still matter full of quality and sensuousness. But to realize 'matter as owned by the

45

bourgeoisie,' it is necessary to eliminate the observer. Since Nature is to be apprehended as it were by a kind of divine apprehension on the part of the observer, in which he stands in no mutually determining relation to Nature, it is necessary to strip matter of all the qualities in which the observer is concerned. Colour, for example. Here the colour involves a subjective element: it is not the thing in itself, but the thing as seen. At first matter is only stripped of colour, sound, 'pushiness,' heat, which all prove to be modes of motion. Motion, length, mass and shape are however believed to be absolutely objective qualities, independent of the observer. However they prove one after the other to be relative to the observer. Thus matter is left finally with no real i.e. non-subjective qualities, except those of number. But number is ideal, and hence objective reality vanishes. Matter has become unknowable.

The categories of Time and Space, regarded as *absolute* categories, express this attempt to remove the bourgeois from active relation with the object. If the object, Nature, can be completely isolated from the subject, Man, it can be expressed in terms of itself—set in an absolute space-time. Man's relation to it is not, in that case, an umbilical cord of mutual dependence; the known Nature is not an active mutually-dependent relation between Man and the rest of reality, but known Nature is Nature absolute and yet in contemplation. This contradiction—a self-sufficient Nature, and yet one contemplated by Man— is the contradiction which drives on the development of physics. Since every quality of Nature is found to contain a subjective element which makes man dependent on something 'out there,' just as it makes the quality dependent on something 'in man,' this contradiction strips all Nature of quality. The most general objective qualities of Nature seem those of Time and Space. Space, the common likeness in phenomena and Time, the unlikeness, seem objective and intrinsic. Surely therefore they are completely qualities of reality-in-itself? Surely man is correct in hypotheticating an absolute framework of absolute Space and absolute Time?

In fact this is a demand that man, the subject, should live out of Time and Space. For if on the one hand we have mind, and on the other hand matter described in terms of itself, then we should have two worlds which do not have anything in common and would therefore be unknowable to each other. But the unknowable does not exist and therefore the closed world of mind and the closed world of physics cannot exist. The famous dualism of matter and mind is something artificially imposed by the special categories of the society which generated philosophy.

Yet this closed world is the aim of bourgeois physics. It is the inevitable presupposition of mechanism.

The characteristic of physics is supposed to be this; it is a world in which each entity is explained by another entity, until you arrive back at the first entity. In *The Nature of the Physical World*, Eddington gives a good picture of this closed world of physics. What is *matter*? Something explicable as a *stress*, which in turn is defined in terms of *potential*, which again is reduced to *interval*, which has to do with *scales*, which are composed of *matter*—and so we have performed the full circle. But, according to Eddington, at this point the reader interjects, 'Please do not explain any more. I happen to

47

know what matter is.' Matter then is 'something that Mr. X. knows,' but for all that Mr. X. remains outside the carefully closed world of physics. Eddington here inserts a diagram of a closed polygon, with Mr. X. *outside* it, though forlornly attached to it.

But if Mr. X. were really outside the charmed circle, how could he come to know it? Mr. X. is in fact Mr. Bourgeois, and it is not his modesty, as might be thought, that keeps him outside, it is his pride. If he comes inside, if Mr. X. is in causal relation with matter, if he *is* matter, he is no longer human desire emerging spontaneously and realizing itself by a mere contemplative knowledge of the mechanical necessities of the object. Such a world must necessarily be a world of the absolute, that is, of an Absolute excluding the human mind as an active part of it. The human mind just wanders on and surveys the frigid scene, without this process of knowing in any way altering it.

Such a world involves the following: There is an absolute Time and Space, independent of the human mind (the observer), in which particles follow absolute paths definable by the Hamiltonian Principle of Least Action. Laplace's Divine Calculator can now come on the scene, and after a lightning glance round in the course of which he grasps the relationship of everything to everything, he can predict the future and thus completely dominate the environment.

12. THE WILD WORLD

We now understand how it is that the Newtonian world presents such a strange likeness to bourgeois society *as the bourgeois envisages it*. It is atomistic. It is composed of individuals who merely proceed on their own right lines doing what the immanent force of each makes necessary. Each particle is spontaneously self-moving. It corresponds to the 'free' bourgeois producer as he imagines himself to be. Events consist of their collisions; and are the product of internal chance.

But a mysterious world force holds all these particles together in one system. Acting as a unifying regulating system, inexplicable and arbitrary, it adjusts, compensates, balances and produces the ordered circulation and self-regulating cycles of the sidereal and solar systems. This corresponds to the bourgeois 'free' market, the law of supply and demand, which holds all the bourgeois producers together, adjusts automatically their relations to each other, and acts as the grand unifying principle of society. It is no accident that this force of gravity is in Newton's mind closely associated with God. The same unconscious forces perturbing and regulating the anarchy of bourgeois society, drive the bourgeois again and again to the altar.

The Newtonian system is of its essence stable and oscillatory. It is like a pendulum. The laws of gravity, of absolute Time and Space, and of the conservation of matter, energy, and momenta, keep the system moving like a pendulum, eternally beating the same path.

But this is precisely true of the bourgeois economic system as the bourgeois economist sees it, in which the market, by virtue of the law of supply and demand, automatically adjusts production to consumption, and price to value, so that there is a perpetual equalization of the needs and production of society, a perpetual

49   E

realization of the greatest possible happiness of the greatest possible number.

Yet we know that in fact the Universe is very different from the stable Universe of Newton. It is a Universe which develops. Solar systems come into being and decay; nebulae condense and grow cold. Life emerges, and grows insurgent and gives birth to consciousness. Mind is born. Hope and despair comes into a world which does not know these qualities. The drama of qualified existence unrolls itself.

But by its very presuppositions, Newtonian physics is forced to deny the reality or relevance of these insurgent 'wild' qualities. They are qualities in which Mr. X. is concerned and therefore unreal. Physics makes a continuous and desperate effort to rid itself of these qualities, but only succeeds in ridding itself even of motion, time and space, its primary categories. Finally it ends up with Lemaître's unstable exploding universe. Even such categories as distance are developing.

And exactly the same is true of bourgeois economics. So far from being a stable society, it is the most violently revolutionary society yet known, continually transforming its own basis and leading to a feverish development of social productivity:

'Constant revolutionizing of production, uninterrupted disturbance of all social conditions, everlasting uncertainty and agitation distinguish the bourgeois epoch from all earlier ones. All fixed, fast-frozen relations, with their train of ancient and venerable prejudices and opinions, are swept away, all new-formed ones become antiquated before they can ossify. . . .'

Crises come with the violence and unexpectedness of earthquakes. Bourgeois society is full of insurgent quality; yet the economists attempt to explain these disturbances, just as the physicist attempts to strip the Universe of quality, by branding them as deviations, as accidents, as mal-functioning, as unreal.

Thus in both cases we have two systems: the system as the bourgeois believes it to be, and as it really is. The first system, the ideal, is subject to the categories of mechanism, i.e. to the characteristics of the machine in society as the bourgeois believes it to be; the second, the real, is subject to the categories of dialectics, of the machine in society as it is known to the proletariat who forms part of it.

But Newtonian physics, with its stable ordered world, is the philosophy of a bourgeois society still stable and not yet embarked on its revolutionary insurgence. It is a society of norms imposed from without; of compromise with the aristocracy. It is the era of manufacture. I have shown elsewhere how this era is reflected in literature of the English eighteenth century, and how it expresses the spirit of bourgeois economy where the market has not yet developed to a stage permitting the machine to become revolutionary, and continually transform its own basis. It is still the machine as hand manufacture, not as the steam-driven factory. It is the machine, only slowly passing out of the era of handicraft and still suffering from a shortage of labour and a restricted market. Presently it will grow insurgent and create the conditions for its own development. It will expropriate the petty bourgeois artisan in thousands and so create its proletariat; it will open up the markets of the world. But at

51

the moment it needs for its slow growth the protection of laws, and labour regulations, which are accepted as norms given by eternal reason and producing a stable 'sensible' society. Hence the mechanistic categories of physics are categories of bourgeois economy in the era of manufacture.

But when the machine breaks loose, and begins to transform society, bourgeois science is also transformed. Other categories grow up beside the older mechanistic categories. The bourgeois class floats to power on the dynamic wave of the machine. The Industrial Revolution has taken place. Man's view of Nature is impregnated with subjectivism, which in bourgeois society is idealism.

Now the bourgeois philosopher sees Nature through rapidly changing economic categories, and hence sees a changing Nature. He sees the change in Nature. Just as a film enables us to see motion in Nature, so does the Industrial Revolution, because of its rapid change of Nature and so of society. The interest of scientists is now directed to change in Nature, and the Darwinian theory emerges, which is a theory of change in Nature explained by the categories of the bourgeois society of the Industrial Revolution, with its *laisser-faire* policy.

Just as the early bourgeois conception of the machine, coupled with the stable categories of manufactured society, led to the development of Newtonian physics, with its stable world and eternal oscillation of bare quantities, the Industrial Revolution, in which the machine produced an instability in society, led to the development of Darwinian biology, in which the categories of mechanism automatically give birth to a progressive evolution of species. The basic relation—the bourgeois separate from the machine—is still the same, but the transformation of society as a result of that relation has led the philosopher to direct his attention to a new field of quality, that of biology or Nature changing.

13. WORLD-VIEW AND SOCIAL CHANGE

And this indicates the way in which the economic categories of a society direct ideology into specific channels. Newtonian physics is not a reflection of bourgeois society; if it were it would not be knowledge about reality; and its practical success indicates its real content of positive knowledge.

Physics is necessarily the science of the most objective components of phenomena; it is the most generalized and formal aspect of matter. It is quantity as bare as possible of quality. As such it is an abstraction. The special circumstances of Newton's day, however, leading to a divorce of the object in contemplation from the active spontaneous subject, made the philosopher imagine that physics was absolutely objective—and thus produced the contradiction whose resolution led to its development.

Thus the categories of bourgeois society directed interest into physics and gave that physics, in addition to its penetrating development, a special distortion—the absolute separation of the object and the picture of a world which was a macrocosm of mercantile bourgeois society.

In the same way society of the Industrial Revolution directed interest into a field of objective quality subject

53

to rapid change; that of biology. It made Man look for change everywhere, and began the development of all the evolutionary sciences: not merely biology, but also geology, cosmogony and the like. This picture of evolution was also given a characteristic distortion.

CHAPTER THREE

MAN AS IDEA

1. THE GENERATION OF IDEALISM

THE Newtonian flourishing of physics was succeeded by the Darwinian growth of evolutionary science. The way for evolutionary theory had already been paved by the development of idealism. Idealism appears in bourgeois philosophy to oppose itself to mechanism, and in a certain sense it does. But if we look into concrete living, we see that both are generated simultaneously. For on the one hand the object, Nature, emerges as the self-contained machine; and on the other hand, as a quite separate phenomenon, Man's desires, his whole activity in so far as this is valued, appears spontaneously, out of the night, and appears to develop of its own, as an independent subject.

Mechanism stripped Nature, the object, of all qualities which had in them any tincture of the subjective, and which therefore made Man dependent on nature. This set free all sensuous active quality as Man's exclusive possession, the attributes of Mind. All the active sensuousness of reality was developed as part of the non-natural science of knowledge. It became a question of thought and thus its development fell to the lot of 'philosophy' —i.e. that part of bourgeois philosophy which, because it is cut off from the object—i.e. from experimental test

—is regarded as the queen of thought and is set above science. It was the peculiar result of the cleavage between subject and object produced by bourgeois economy that the sensuous active element in concrete living was developed separately from science as idealism.

Berkeley, Hume, Kant, Fichte, Schelling and Hegel represent the stages by which the subject is cut completely free from the object. I do not propose to deal in detail with these stages here, as they will be familiar to the student of philosophy and to non-students would be too technical for brief explanation. The point is that this was a process in which man or mind, figuring as active, sensuous subjectivity, was stripped of all those qualities which had an objective component in them. But since no quality emerging as a phenomenon is absolutely objective or subjective, no quality is situated in an absolute self-sufficient Space and Time, nor does any quality exist completely out of Space or Time. The stripping from subjectivity of all qualities containing objective components left it as bare as matter when it was stripped of all subjective quality. Matter was left with nothing but mathematics existing in the human head. Subjectivity was left with nothing but the Idea; and obviously this could not be the idea in the material human brain, for this would tie the Idea to matter. Hence this final reality was the Idea existing out of the human brain—the Hegelian Absolute Idea.

Not only has subjectivity by then been stripped of activity and sucked of sensuous blood, but it has in fact become objectivity, for the Idea existing apart from the brain is objective reality and therefore enters the category of matter. Idealism has become materialism, just as mechanical materialism when it ended as mathematics, had become idealism. Mechanism and idealism, although they seem irreconcilable opposites, are only so in the sense that they are different sides of the same penny. They are produced by the cleavage of subject and object which results from the special conditions of bourgeois economy.

2. THE ADVANCE TO DIALECTICS

It therefore fell to bourgeois philosophy to develop in a classic way the active sensuous side of existence. Now what is active is changing and thus the development of subjectivity was the development of an evolutionary philosophy. This became evident with the emergence of the Hegelian Dialectic, which is an evolutionary theory of subjectivity. The categories of mind here generate each other in an evolutionary way. Thought has become full of history and time.

But we saw that this was a subjectivity carefully pruned of the objective component. Hence it was a subjectivity whose activity was an activity on nothing real—on 'appearance'—which is how the object figures in the realm of mere experience. Hence subjectivity in the form of bourgeois philosophy lacks the essential test of practice, and experiment. It lacks activity upon the object which of course can only be realized in practice. Yet in fact all subjectivity, even mere knowing, is activity through objects. Hence subjectivity strangles itself.

None the less subjectivity, by gathering into itself all the sensuous active qualities of existence, does, even when robbed of the object, contain the impress of

57

material reality, like bark stripped from a tree. By comparing these unanchored qualities among themselves, it is possible in a confused way to extract the most general laws of activity, and change, just as by comparing the categories of objectivity among themselves one gets the confused but general physical laws of mechanism. These laws of subjectivity are laws of logic. It is not formal logic, but the Hegelian logic of dialectics.

Dialectics, as developed by Hegel, does not therefore merely express the laws of 'thought.' Because the 'thought' of Hegel is really subjectivity or active sensuous existence, in the widest sense, Hegelian dialectics attempts to realize the most general law of change. It grasps at the emerging of the unlike, the birth of quality, the movement of evolution, the passing of history, the process of real Time. In this it proves the opposite of mechanism. Mechanism is concerned with the persistence of matter, the conditions of stability, the survival of the like, the shuffling of quantity, of the substance below change, the isotropic framework of space.

But dialectics can only be filled with content by activity upon the object—that is, by practice and experiment. Since the object did not exist for Hegel, his dialectic could never be filled with realistic content, and remained a beautiful and intricate mill grinding the air of theory and producing nothing but his prejudices and aspirations.

3. THE EXPLOSION OF THEORY

The failure of Hegel was inevitable. Because of the bourgeois conception of the machine and of the general make-up of society, human desires include all the blood-warm valued qualities of existence and emerge spontaneously on the scene—their past history is veiled in the shadow of the free market. It is not possible to see the process by which they were originally determined through the 'production complex' at whose heart lies the machine. Hence mind seems undetermined—that is, spontaneous and 'free.'

But this reflected separation of theory from practice and of desire from object is a reflex of a cleavage of classes which is fundamental to society. On the one hand there is the bourgeoisie, in whose heads the theory of society is concentrated, by virtue of the class division which has given to the bourgeoisie the task of the conscious supervision of the labour process. On the other hand there is the proletariat who actually deal with the object, Nature, but to whom theory is a 'reserved' item, a privilege of their betters. The philosophy of the bourgeoisie sunders theory from practice because they are sundered in the concrete living of society.

The study of the object becomes the study of the object *in contemplation* and therefore lacks the dynamic reality of struggle. The effect of this is to make science too mechanistic and rob it of living theory. That is not to say science has no theory: it is impossible to have any practice without a theory, but science's theory is the minimum theory possible, a theory which is empiricist and opportunist because it springs directly from practice. It is not a theory which has been evolved to meet the needs of a man's whole life in society—including his scientific speculation. It is a specialized theory designed only to meet the needs of a man as a scientist and not as a man with blood in his veins who must eat, labour, marry and die. This limitation is pointed out with pride by modern scientists. It leaves room for God, they explain.

Take the case of physics. There is first the general theory or philosophy of mechanism, which the bourgeois scientist adopts unconsciously. He has no idea it is a metaphysics: he imagines it to be the only way of looking at things scientifically—i.e. objectively. He supposes that the object as it appears in bourgeois economy is the only way in which Nature can appear to men. This philosophy is common to all sciences.

In addition there is the specialized theory springing directly from the practice of physics, which from time to time contradict this theory and leads to its improvement.

All goes well till a point is reached where practice with its specialized theory has in each department so contradicted the general unformulated theory of science as a whole that in fact the whole philosophy of mechanism explodes. Biology, physics, psychology, anthropology, and chemistry, find their empirical discoveries too great a strain for the general unconscious theory of science, and science dissolves into fragments. Scientists despair of a general theory of science and take refuge in empiricism, in which all attempt at a general world view is given up; or in eclecticism, in which all the specialized theories are lumped together to make a patchwork world-view without an attempt to integrate them, or in specialization, in which all the world is reduced to the particular specialized theory of the science with which the theorist is practically concerned. In any case, science dissolves in anarchy; and man for the first time despairs of gaining from it any positive knowledge of reality.

This is the state of bourgeois science at the present day, and the crisis in physics is only a special expression of it. And of course it is only a still wider version of the general movement of bourgeois economy into anarchy; the productive forces at all points have expanded and burst the confines of the productive relations. Humanized nature seems to be escaping, like a Frankenstein monster, from the meshes of naturalized men. The machine is getting out of control of the mechanic. This points the way forward. The disintegrating old contains the developing new. A new set of productive relations; a new society; new ideological categories; a new or wider world view.

4. REALITY AS APPEARANCE

But the effect of this disruption of the old bourgeois world-view on scientists is to throw them back, for an explanation of reality, upon those qualities, in all their active sensuousness, which they had successively abandoned to subjectivity. We saw that the development of subjectivity fell to the lot of so-called philosophers.[1] I say so-called, for while they were certainly philosophers, their claim to regard their field, subjectivity, as all philosophy, is untenable. Philosophy can only mean the most general theory of practices, and therefore it must include the theory of science. But philosophy merely concerned itself with subjectivity which excluded even mind regarded as an object (i.e. experimental psychology). It was an important moment for so-called philosophy when psychology slipped out of its grasp into the camp of the

[1] Also artists, but this is another story.

61

experimentalists, for this finally exposed the completely anchorless state of its ship. It was subjective activity, *active upon nothing at all.*

Hence the feature of the present crisis in physics, is that the 'scientists turn to philosophy.' What in fact this really means is that they find their philosophy of mechanism shattered beyond repair by the progress of science and turn to the other side of the medal, to the erstwhile schismatics of subjectivity, to fill the breach. Scientists now seek in the 'laws of thought' a certitude which they cannot find in the laws of the object.

But we saw that the subjectivist had in the interim developed on the same lines as the mechanist. He had stripped the subject of all objective qualities until nothing was left but the absolute Idea—the Idea existing objectively out of the brains of men. But in doing this he had stripped subjectivity of the subject—man. Hence when the mechanist turned to the subjectivist for assistance he found that the subject had vanished. The object had for the mechanist become unknowable, or a thing-in-itself, or had ceased to exist—these are all different ways of putting the same discovery—and now he found that exactly the same has happened to the subject.

What then could exist, *philosophically*, for the scientist? Only phenomena—that is, appearance—the conscious field regarded as independent of subject and object. The subject-object relation is regarded as existing apart from its terms. This has some resemblance to the absolute idealism of Hegel, but because the scientist regards even subjectivity mechanically, he cannot accept the dialectic logic of Hegel. Hegel's dialectics ostensibly draws its validity from the power of reason. It rests on the inward and unquestionable witness of the 'I' which thus, in the alienation of the Absolute Idea, appeals to itself to deny itself. But the scientist, by his training and experience, cannot accept the 'I' as the criterion of validity. He is born in practice. Hence he cannot accept the subjective authority of the Hegelian dialectus. He can only accept phenomena as they come. This is positivism.

But in fact phenomena emerge from the concrete living of society, and this is an active struggle of Man and Nature. If Man and Nature are ruled out as unreal and non-existent, phenomena all have absolutely equal validity: hallucination and real perception, scientific theory and barbarous logic, there is no means of choosing between them. Truth is meaningless. We are in fact—if positivism is carried out logically—back at the subjective idealism of Berkeley and the scepticism of Hume. Positivism is solipsism. Nothing exists but my experience.

But in fact the positivist will not face up to his premises. He continually smuggles in some co-ordinating principle which in fact presupposes the existence of the very things he cannot prove. For example, he includes in phenomena 'other people's phenomena' and so accepts the findings of science and other organized knowledge. Yet in fact he has no right to accept other people's perceptions except by admitting the link, his human brain and other human brains, which means admitting the subject, Man, and the object, the matter of which brains are composed. He smuggles in 'principles of economy' which are simply logical laws admitting therefore the validity of the subject; and 'laws of efficiency' which admit the existence of the object through the test of the practice.

Mechanism sacrifices theory to practice. Subjectivism

sacrifices practice to theory. Positivism denies the validity of both, but in fact is always driven to smuggle one or other in by a back-door, because the very reason for its existence is that theory has been whittled away by mechanism and practice by subjectivism. Hence positivism is always a confused, amateurish and dishonest philosophy. It makes a degradation of bourgeois thought as compared to the simple grandeur of Newtonian physics and the world-dominating insurgence of Hegelian dialectics. This confusion is very clear in the writers of the older positivists, Mach and Pearson, and the newer positivists, Eddington and Jeans. Their writing is full of contradictions, they shift from one premise to another without realizing it: their writing is a mesh of excluded middles and *non sequiturs*, directly it deals with philosophical questions.

5. THE SCREEN OF PHENOMENA

'Sensation is nothing but a direct connection of the mind with the external world; it is the transformation of energy of external excitation into a mental state . . . the sophistry of idealist philosophy consists in that it takes sensation, not as a connection of the mind with the other world, but as a screen, as a wall which separates the mind from the outer world.' (Lenin.)

Consciousness (phenomena) is a relation between Man and Nature, but positivism attempts to take the relation without the terms. This in itself is a result of the splitting of the terms in concrete living.

So split, consciousness, part of the subject-object (the

'theory' of it) ceases to be active. It is impossible to have real activity without two terms, without a contradiction, and a unity of opposites whose activity springs from their interpenetration. Hence consciousness becomes a mere passive 'reflection' of the world; its function becomes merely to be a pale copy of existing practice. The relation of knowing ceases to be an active and mutually determining relation, and becomes a godlike apprehension separate from material reality. But directly it is cut off in this way, it loses its real content.

Hence ideology in bourgeois society becomes distorted to a mere symbol or code-word for reality. Reality knocks on the nerve endings and these are 'interpreted' as consciousness by the subject. This theory of consciousness as mere reflection leads to a regretful admission that it is a 'misleading' reflection. For since all the known subjective qualities (colour, scent, shape, mass, pushiness, beauty) are merely symbolic ciphers for the thing in itself the 'reality' codified is a queer grotesque spectre, built vaguely out of the most objective qualities obtainable. Thus according to Eddington, the real table is a swarm of molecules buzzing hither and thither, and is totally different from the table we see. The table we see is a mere fiction, a symbol of the real thing. Consciousness here has become a screen. Hence the severance of the subject and object, of Man's natural desires from nature as known by Man, leads to a splitting of consciousnesses. The consciousness of the bourgeois philosopher is torn into two. One half of it flies to the objective pole, to become a bare 'copy' of practice on the object and so eventually come to a stage where the object seems unknowable by consciousness.

Moreover, because practice advances on different fronts, this theory splits into several theories adhering to different practices (biology, physics, psychology, etc.). The other half flies to the subjective pole, to become a 'spontaneous' undetermined desire. This emerges as mysticism and religion, with a subject as unknowable as the object. This double decadence into positivism and mysticism is clearly shown in the following quotations from Eddington:

'In regard to our experience of the physical world, we have very much misunderstood the meaning of our sensations. It has been the task of science to discover that things are very different from what they seem. But we do not pluck our eyes out because they persist in deluding us with fanciful colourings instead of giving us the plain truth about wave-length. It is in the midst of such misrepresentations of environment (if you must call them so) that we have to live. . . . In our scientific chapters we have seen how the mind must be regarded as dictating the course of world-building; without it there is but formless chaos. It is the aim of physical science, so far as its scope extends, to lay bare the fundamental structure underlying the world; but science has also to explain if it can, or else humbly to accept, the fact that from this world have arisen minds capable of transmitting the bare structure into the richness of our experience. It is not misrepresentation but rather achievement—the result perhaps of long ages of biological evolution—that we should have fashioned a familiar world out of the crude basis. It is a fulfilment of the purpose of man's nature. If likewise the spiritual world has been transmuted by a religious colour beyond anything implied in its bare external qualities, it may be allowable to assert with equal conviction that this is not misrepresentation but the achievement of a divine element in man's nature. . . .

'. . . We have to build the spiritual world out of symbols taken from our own personality, as we build the scientific world out of the metrical symbols of the mathematician.'

'. . . The idea of a universal mind or Logos would be, I think, a fairly plausible inference from the present state of scientific theory; at least it is in harmony with it. . . .'

'. . . The materialist who is convinced that all phenomena arise from electrons and quanta and the like controlled by mathematical formulae, must presumably hold the belief that his wife is a rather elaborate differential equation; but he is probably tactful enough not to obtrude this opinion in domestic life. If this kind of scientific dissection is felt to be inadequate and irrelevant in ordinary personal relationships, it is surely out of place in the most personal relationship of all—that of the human soul to the divine spirit.'

'. . . The physicist is not conscious of any disloyalty to truth on occasions when his sense of proportion tells him to regard a plank as continuous material, well knowing that it is ''really'' empty space containing sparsely scattered electric charges. And the deepest philosophical researches as to the nature of Deity may give a conception equally out of proportion for daily life; so that we should rather employ a conception that was unfolded nearly two thousand years ago.'

'. . . Starting from aether, electrons and other physical machinery we cannot reach conscious man and render count of what is apprehended in his consciousness. . . .'

67

'. . . If those who hold that there must be a physical basis for everything hold that these mystical views are nonsense, we may ask—what then is the physical basis of nonsense?'

'. . . We have associated consciousness with a background untouched in the physical survey of the world and have given the physicist a domain where he can go round in cycles without ever encountering anything to bring a blush to his cheek.'

'. . . The conclusion to be drawn from these arguments is, that religion first became possible for a reasonable scientific man about the year 1927.'

'Heaven is nowhere in space, but it is in time. . . . Science and theology can make what mistakes they please, provided that they make them *in their own territory*; they cannot quarrel if they keep to their own. . . .'

These quotations, taken at random from the final chapters of Eddington's book, indicate the extraordinary confusion and helplessness of the scientists of to-day, faced with the break-up of the old bourgeois world-view. On the one hand objectivity, Nature, has become a game, a symbolism, a separate domain where the physicist can go round in cycles without encountering anything real. Nature has become unknowable.

And Man, the subject, dragging with him all the rich qualities of interesting life, has entered the arid regions of *theology*. Could reaction go farther? Because physics has made of Nature something no one can believe as real (a swarm of sparsely distributed electric charges) it is no longer necessary to believe in the refined 'unitarianism' of modern Broad Churchmen—we can go right back to the Virgin Birth, the miracles of the loaves and fishes, and the 'simplicities' of the New Testament narrative. The wild Elizabethan human desires set free by the bourgeois market have become pious. The machine planned by the bourgeois to satisfy his wants has become unknowable; it has slipped out of his grasp down into the night of the proletariat.

6. THE RE-DISCOVERY OF THE OBJECT

For, in fact, this is where Nature has disappeared. The severance of subject from object by the development of a class cleavage in society, has resulted in that part of society which groups itself round the machine, becoming increasingly organized Man—Man organized by Nature. It follows the grain of objective reality and enters increasingly into the production complex of humanized nature. This group in practical contact with Nature is increasingly *proletarian* society—society debarred from consciousness by the conditions of its existence. It is active of Nature in a blind way—but it is active. It is true that in the experiments of physics for example the bourgeois is in active contact with Nature, but only on a small front. Even that contact is enough, as we have seen, to produce a disintegration of his whole world-view.

But in the main the most important part of objective activity is handled by the proletariat. The most elaborate and intricate organizations produced by the incursion of Nature into society and the humanization of Nature as a result of the division of labour are organizations of the lives of the proletariat. The dizzy unfolding of Nature within society which is modern civilization takes place within the boundary of the proletariat. The bourgeoisie rides on top of this terrific pregnancy, unorganized except in the old State forms and these forms become increasingly arbitrary, increasingly the product of the apparently blind desires of the bourgeoisie.

They stand in a coercive one way *owning* relation to the forces wielded by the proletariat, and therefore seem all the more free of the object, and masters of Nature. But in fact the object has now retired completely into the night of the exploited class. The bourgeois ignorance of the object, and of the determining relation it has over their lives, makes them its slaves, tossed hither and thither by slump and boom. By cutting finally the cord that binds their desires to the necessity of the object, and making desire and subjectivity a matter of faith and theology, the bourgeoisie prepare the ground for their ejection from power. The pregnancy of the proletariat with the humanized object is a pregnancy which can only issue in revolution.

We saw that practice must inevitably carry with it some theory, however partial and specialized—a theory perhaps distorting and negating the general world-view of the practician. Thus the practice of the physicist carries with it a limited and bloodless theory which conflicts with the older bourgeois world-view and produces a helpless dualism or anarchy. In the same way the actual experience of the proletariat produces a special theory of its own, the theory which springs from the practice of trade union organization.

This limited theory is directly contradictory to the whole theory of bourgeois society, in which freedom lies in absence of restraints, and in a completely free market for labour-power and wages. Trade union organization, with its restrictions and limitations, on labour, negates this basic consciousness of bourgeois society, but it is forced on the proletariat by the necessities of concrete living. Hence it has a shattering effect on such portions of bourgeois consciousness and world-view as have been implanted into the proletariat.

None the less the plentitude of freedom and therefore of consciousness still remains in the sphere of the bourgeoisie. The proletariat, alone, cannot rise beyond trade union consciousness. This consciousness, although it sees freedom to be the outcome of restrictions of the market and thus denies bourgeois ideology, yet proposes a freedom which is dependent upon the existence of a bourgeoisie, a freedom within the pores of bourgeois society. It is thus a consciousness limited on every side by bourgeois consciousness and unable to make itself independent, unable to advance to the status of a new world-view.

But the progress of capitalism transforms its own basis and creates conditions of unfreedom even for its own bourgeoisie. The big bourgeoisie grows and expropriates the smaller, who is forcibly proletarianized; or else the big bourgeois forms an alliance with the feudal aristocracy to prevent the advance of the other section. Thus a section of the bourgeoisie is driven to ally itself with the proletariat. Part of this section have no other aim than to use the power of the proletariat to wring concessions from the big bourgeoisie and bring back the old conditions of existence more favourable to petty bourgeois ideals, conditions of existence in which a petty bourgeois could flourish without danger from monopoly capital. This gives rise to the movements of anarchism and

71

reformist social-democracy, which remains within the categories of the bourgeois world-view and try to drag the proletariat into it.

But 'when the class struggle nears its decisive hour, the process of dissolution going on within the ruling class, in fact within the whole range of old society, assumes such a violent, glaring character that a small section of the ruling class cuts itself adrift and joins the revolutionary class, the class that holds the future in its hands . . . a portion of the bourgeoisie goes over to the proletariat, and, in particular, a portion of the bourgeois ideologists, who have raised themselves to the level of comprehending theoretically the historical movement as a whole. . . .'

This small portion joins the proletariat. It does not attempt to use it as a tool to fulfil its own desires because it has been forced in practice to comprehend the historical movement as a whole—i.e. the victory of the proletariat and the impossibility of a return to petty bourgeois ideals. It does not however shed its bourgeois consciousness but drags this with it into the proletariat. The object has already slipped out of the grasp of the bourgeoisie as part of its world-view. The subject has however developed to reach its climax as the Hegelian dialectic. This is a moving dialectic, and it is within the framework of the Hegelian dialectic that this section of the bourgeoisie comprehends the historical movement as a whole at the same time as material causes drive it to a revolt against the existing system and an alliance with the proletariat.

We have already seen that the object, Nature, in its full development by capitalist society, had disappeared into the concrete living of the proletariat. This class was pregnant with Nature as increasingly realized in society by the division of labour. Hence when bourgeois subjectivity in the shape of its most advanced development, dialectics, is driven by material conditions into the bosom of the proletariat it once more encounters the object, and the object is now as a result of technicological advance, in its most highly humanized form. We saw that dialectics, in spite of its logical rigour and world-embracing grandeur, became mere mystical mumbo-jumbo because it was subjectivity active upon nothing, upon mere appearance and remaining therefore unfounded theory. But in the heart of the proletariat it encounters the object.

It must not be thought that this is a kind of marriage of long-separate twin souls who suddenly embrace. It is not a case for example of bourgeois mechanism (objectivity) being fused with bourgeois idealism (subjectivity). For mechanism loses the object ultimately without developing the subject, and dialectics ultimately loses the subject without developing the object. Materialism becomes idealism and idealism materialism. Their fusion therefore produces only positivism—the relation without the terms. This was bound to happen because one started with a *contemplated* object and the other with a *spontaneous* subject. Hegelian dialectics cannot marry the object, wrapped in the proletarian night, in the world of theory, for the object is not yet conscious. The object is wrapped in night, and subject and object live in different worlds. Before the marriage can take place, the object must be made conscious by activity, by practice upon it in a world-changing way. It is not a mere case of 'fitting' the results of science into the categories of dialectics.

73

Dialectics must become active upon the object in real life; only in this way can dialectics become full of content. And since the object is at first entirely concrete and unconscious, this abstraction must begin in the least abstract and most practical way, by making the proletariat conscious of its most general class interests and goal, and by developing the theory of the proletariat from that primary and fundamental activity.

For this reason dialectics became with Marx and Engels a practical revolutionary theory, and it is in this way, as the result of practice, that it becomes dialectical *materialism*. From this most concrete basis, dialectical materialism can then proceed to draw in the ideological products of society—the sciences, ethers, art—and reform them within the new categories.

Can dialectical materialism escape in its development the limitations of bourgeois society, in which the subject became separated from the object? The class of which it has become the world-view, the proletariat, is pregnant with the object and this has produced an increasing organization, a revolutionary expansion, which will continue until the proletariat has become a whole and thus has realized a classless society. As this expansion takes place the revolutionary class, pregnant with the object, sucks more and more of the subjectivity, the consciousness of society, into its sphere. And thus as it actively expands, as scientists, artists, and 'philosophers' desert the bourgeois class and enter it, its world-view, dialectical materialism, synthesizes more and more of the genuine but anarchic and dispersed elements of bourgeois consciousness. But this new consciousness is not one in which active subject is parted from contemplated object, and

74

the real activity of society sinks into the night of an unconscious class. In dialectical materialism subject is restored to object because in the society which generates it, consciousness is restored to activity and theory to practice.

CHAPTER FOUR

THE DISTORTION OF PHILOSOPHY

1. THE MONOPOLY OF CONSCIOUSNESS

THE essence of the distortion of ideology by a class society is this: a class society consists of a ruling class and an exploited class, and the consciousness of society is the consciousness of the ruling class. This follows from the very mechanism of class formation. At a certain stage division of labour demands that certain men stand to social production in the relation of supervisors, managers and overseers of labour. Only by this means can productivity advance to a higher stage.

Chosen in the first place by tribal society for innate qualities of intelligence, the process of development changes them from overseers or custodians of the means of production on behalf of the community, to owners of them in their own right. But the higher consciousness necessary for the supervisory role persists with this class in their development, and thus the whole consciousness of society gathers at the pole of the owning class.

As the obligation of supervision becomes more and more an absolute right of ownership, the practice passes more and more out of the orbit of the ruling class, who handle increasingly only the theory of society. They stand outside the main organic complex in which Man struggles with Nature. Practice, divorced from theory, yet secretes,

76

as it must, a theory in its pores which represents only a limited and objective consciousness, contrasting with the formal and unanchored consciousness which is the seat of privilege.

The former is a practical but departmentalized theory, split into hundreds of units. The latter is generalized, but is theoretical and stamped through and through with class illusions.

Hence in a class society, the consciousness developed by society receives a characteristic distortion due to the fact that theory is sundered from practice in a special way, and only a part of concrete living falls within its scope. The rest passes out into the night of the other class and returns again transformed—no one knows exactly how.

This is not to say that a classless society is in possession of absolute truth. The classlessness of primitive tribal communism is very far from being the recipe for absolute truth. On the contrary the conscious theory of such a society is primitive, poor in content, and undifferentiated, consonant with their primitive state of society. For Truth—i.e. the living theory of a society—is not an absolute good dropped from heaven; it is an economic product. It is a specific penetration of Nature by Man, and Man by Nature which, as by a mutual reflexive movement, has given rise to an image of each in the terms of the other. In the theory of consciousness, Man interprets Nature in images of himself. In the practice of production Nature is minted in human metal. But science, art and ethics—the vehicles of theory about society—are generated by the development of society itself. The glittering superstructure can only rise upon the foundations of economic production on which it acts and re-acts. It is

77

this action and reaction which produces the continual modification of the superstructure by the base, like the nourishment of a power by its root. There is a rising and a falling current of sap. Theory is negated by practice, and modified accordingly; the new theory opens the way for a more effective practice.

In a class society however there is a characteristic splitting between theory and practice which isolates the superstructure to an increasing extent. At the same time this distortion of the superstructure is the result of a growth and rearrangement of the root system, which makes for a more efficient economic production and hence—at first—for an elaboration of the superstructure. It becomes specialized, and blossoms. It is only when the specialization and root development passes a certain stage that the life of the whole organic structure is affected.

2. PRIMITIVE MATERIALISM

In this respect therefore a distinction must be made between the special *distortion* of ideology by a class structure in society and the limitations imposed on ideology by a given system of production. These two factors do not necessarily work together and a specific distortion by overcoming the limitations of production, may give rise to a luxuriant though one-sided growth.

Primitive ideology conceives reality free from the distortions of a class society. It is materialistic in its outlook. Animism is primitive materialism. Class society completely separates mind from matter, and the activity of the Universe from the stuff which is active. It separates growth and change, as an immanent force, from that which grows and changes. It analyses motion into space, the most generalized form of the persistent object, and time, the most abstract form of phenomenal activity. Theory lies apart from practice.

When idealism finds that the savage makes no such distinction, but observes a world in which things move because of an immanent power, and change by virtue of inner activity, not being changed or impelled on by forces outside the Universe, it at once imagines that the savage has first separated mind, subjectivity, from objects and then thrown them back again into the objects in the form of Mana, Oronda, spirit, or power. But of course the savage has never been through this prior stage. The savage sees the world (or rather objects, he does not yet see a Universe) as self-changing and self-developing things. He himself is a self-changing, self-developing thing and therefore he makes the mistake of supposing that the activity of objects is to them what subjectively his own is to him. He attributes to them will, feeling and desire, as he knows these things. Perhaps to this degree he is guilty of 'animism.' But he is by no means guilty of the animism ascribed to him by class society, that of throwing back into nature all the categories of subjectivity, of spirit, sucked from it by a class ideology.

To this degree dialectical materialism is a return to primitive materialism just as communism is a return to primitive communism.

It returns to reality, the life that has been extracted from it by the class distortion of ideology. But it is materialism gathering up into itself all the richness of ideological history which has taken place as a result of

79

the economic development which this cleavage made possible.

It synthesizes mechanism—the bourgeois development of objectivity and practice—with its sundered pole of subjective theory. And in exactly the same way, on the economic level, it gathers up into itself, and resolves the contradictions, between the organization made necessary by the division of labour and the personal freedom, made possible by the plenty which is the result of division of labour.

Primitive materialism is materialist because it gives to substance an inner activity and capacity for history which is abstracted from it by class society. Of course it is a crude self-moving power—that of Mana. In the same way primitive materialism is monist—it ascribes to all things a sympathetic influence on others, like the universal law of gravity of Newton. This 'sympathy' is also crude and subjective—that of magic.

3. THE COSMIC MARKET

Bourgeois practice also gives objects a certain self-moving power, under the abstract guise of force, and a certain monism under the guise of gravity or the space-time continuum. In this unity of pluralism and monism it is still materialist. But already ideology has been robbed of all the sensuous and qualitied richness which is present in reality in however crude a form in primitive material-ism. These have been delivered over to theory for a separate development.

The abstraction of bourgeois objectivity is due to a similar abstraction in society. The growth of the market

80

equates all commodities to a common denominated—
exchange-value. Even men are reduced to a common
labour-power. This ruthless stripping of all qualities to
an abstract commonness reaches in the era of manufacture a limit which gives us on the one hand the abstract
'economic' Man, the producer, the common unit of
labour power and on the other hand the common regulating principle, the market.

This is reflected in Newtonian physics. The particles,
independent and self-sufficient, travelling on ideal right
lines at equal speeds (except for their collisions) are
equivalent to the abstract producers, the units of society.
And the law of gravity represents the regulative principle
of the market.

How impossible it is for bourgeois man to escape at
that period from this conception of objectivity is shown
by the contemporary conception of Leibniz, superficially
different from Newtonianism but in fact the same in
essence. The self-moving particles are the monads. The
regulating principle of the market is the God to which
all the monads open their windows.

The law of gravity, God, and the 'free' regulative
market are conceptions that fit in with bourgeois society
like hand in glove. The abstract God of bourgeois deism
is a God peculiar to a society which has reduced the
mysterious unknowable element in society to the market
which equates everything to a common term, exchange-
value. God is always humanity's name for its confused
perception of the part of society which is hidden from it.
To the primitive the divine elements is above all magic
and magic is 'sympathy.' It is the mysterious but undifferentiated instinct which binds together a tribe. It is the

herd instinct realized in an economic shape. As such it is one category of the interconnection or determinism which everywhere secretly unites phenomena and makes the Universe one. The primitive projects his subjective experience of this interconnectedness, which is nothing but tribal solidarity, into the object, and sees Nature as united by magic—a 'feeling' between things. And as division of labour occurs and individuals realize themselves in the tribe as elders, chiefs and kings, he sees this magic wielded by individuals, who presently acquire the status of Gods.

Bourgeois practice becomes stripped of quality and human warm-bloodedness. Magic now changes into determinism, in which the possibilities of man's active charge of matter are generalized in a framework of causality. Because this is free from the savage's subjectivity, it is real scientific causality and not the 'feeling' of magic. But this very abstraction from subjectivity has robbed determinism of all quality. God becomes a kind of self-moving necessity. Hence bourgeois man's confused notion of what unites society appears as the bourgeois God, the monotheistic principle of existence.

4. GOD-MAKING

But by reducing quality and use-value to exchange-value, bourgeois society could not make quality disappear. It could only prise quality loose from the object to float round as subjectivity, as 'spontaneous' human desire. All this subjectivity is developed separately, it also is a confused perception of society but of the opposite face; it is also attached to God. Thus the God of bourgeois society is a compound God, playing a dual role. On the one hand he is God the colourless abstract principle of theology, source of necessity and law, the God of Malebranche's and Descartes' philosophies. Such a God is unappetizing. He is a symbol for man's faith in practice which, because it is torn apart from theory, is a faith *in* practice and not a theory of practice. On the other hand he is God the focus of subjectivity and quality, the God of the mystics, the God of the Trinity; the personal God: the God to whom it is possible to ascribe the Virgin Birth, the Crucifixion, anger, and an interest in the individual. Here we have an appetizing God who is the same as the other. He is a *human* God, just as the other is a natural God. This confounding of Gods is the source of all the contradictions of religion—why does a kind God allow us to become lepers, and children to be hurt, for example? The God full of human values who is yet forced to permit evil because he is caught in the wheel of his own infinite justice and respect for law is a reflection of human desires trapped by natural necessity because the interplay of the two is not yet understood.

But it is wrong to suppose that these two different Gods could have been fused—that theology and mysticism could have come to terms. For their fusion would mean the reunion of theory and practice and therefore the disappearance of the confusion which led to God. The divine principle of the savage, the unifying magic which is also causality, does not suffer from this cleavage because it is a causality full of feeling and warm blood. They fly apart in a class society and it is precisely their flying apart which develops on the one hand causality and on the other hand subjectivity.

83

The theological God represents as it were the back parts or wounded stub of objectivity or practice. Theory appears to be ripped from practice and objectivity is cut off from subjectivity in consciousness, because the shadow of the night of the exploited class lies over their connections and makes them secret. The personal God is the mutilated end of subjectivity. Yet if they could be fused, if the underground connections between objectivity and subjectivity could be dragged to light (because the exploited class has come into the possession of consciousness) then everything would be plain, and there would be no need to give the mysterious name God to a clearly revealed process of society.

This separation of theory from practice in society, which gave rise to the God of class society with his special dual role of abstract monistic law and human quality, reflected a division of labour which was a necessary stage of evolution if productivity was to advance. It was therefore the means of advancing scientific thought and human feeling. Logic, as with the schoolmen, and poetry, as with the Greek tragedians, were tied to complexes of thought whose lineaments, bathed in the penumbra of a class society, necessarily took on a Divine mien. God was the scaffolding of an undeveloped consciousness.

This dual role of Deity is not peculiar to bourgeois philosophy. It is general for all class society, in which a cleavage between theory and practice must necessarily take place. In all the developed religions we see a monistic abstract tendency which is monotheism in embryo. Even in the most fantastically pluralistic pantheons of Egypt or India, this abstract God appears as necessity, or the

84

Divine Principle, as Brahma, Karma. It is the Law, to which even the Gods themselves are subject: and it is expressed also in the thought that all the Gods are aspects of one personality. In this way man expresses his confused perception of science. He has a formal hypothesis for his nascent understanding of the interconnection of everything, as this interconnection is coming to light in the practical exploration of Nature by society.

But this interconnectedness is denied by the cleavage in society, which wrests subjectivity from objectivity. Hence subjectivity appears in the manifold guise of the Gods with all their rich personalities and endearing or formidable traits. Man thus exercises an aesthetic function which has been confused by his role in society. This is the realm of mythology as opposed to art. Man exhibits capacity for making false concrete images which yet express real subjective truths.

This twin division has sprung from magic for which the world is full of interconnectedness and self-motion but only as qualities of *feeling* and therefore correspondingly crude and simple.

Bourgeois philosophy expresses this contrast between monotheism and pluralism in the sharpest way. On the one hand the refined theological concepts of Hegelian philosophy, in which God becomes a depersonalized Idea like that of gravity; on the other hand the preservation of all the barbarous mythology of early Christianity because of its warm human quality. These myths lose the fluidity of legend and become fixed, like a piece of journalism. But what signals that this marks the *final* stage of religion, and the oncoming of a classless society in which these penumbras will not be cast, is that both

85

abstract and personal Gods are fossil Gods. In the bourgeois era religion loses its artistic myth-creating power and merely preserves the myths and hagiography of the classical and medieval eras: and equally theology cannot escape from musty Platonism and scholastic reasoning. The life has gone out of both, and this life reappears elsewhere as science and art of an unprecedented luxuriance, even though both the science and art are still distorted by the necessities of appearing in a class society, and cast a shadow in which mysticism is bred.

5. NATURE AND THE SLAVE-OWNER

The distortion of the world-view by its generation in a class society varies with the basis on which class divisions rest. In bourgeois society the freedom from social 'restraints' which is its form produces on the one hand an active but 'spontaneous' subjectivity, and on the other hand, a merely contemplated necessity.

In Aristotelian society, however, the distortion is of a different form. There is no longer a cosmos of self-moving particles whose movements are automatically regulated on a universal scale by a mysterious force of gravity; nor does subjectivity appear as something completely alien and spontaneous; for the free commodity market, of which such a world-view is the reflection, is not fully developed in such a society. Pre-bourgeois class society is a slave-owning or serf-owning society, and hence its philosophy of objective reality is one of coercion or will; coercion is not veiled in such society, for production is openly determined by the will of the slave-owner. True, coercion is the moving principle of bourgeois society, but it is veiled and not conscious, and men see reality through a glass coloured by their subjective relation to society. But coercion is conscious in earlier class society. There is no free market into which the exploited toiler can bring his labour-power on a spurious basis of equality with the owners of the means of production; nor does the social will emerge from the market in a spontaneous abstract way like a force of Nature. Social will is the lord's will—the slave produces directly for him—and the relation is simple and coercive. The master proposes an end and the slave fulfils it as a matter of coercion. Thus the physical world viewed by the ruling class of such a society is a world of ends and purposes —it is teleological instead of mechanist. Or rather it is mechanistic in a slave-owning way. It still obeys the categories of mechanism for these are merely the categories of objectivity but now the machine is a slave-owning machine and not a capitalist machine.

The conscious relation of the ruling class to men engaged in changing Nature to meet social desire is different, therefore Nature looks different to them. They explore it with a different microscope. The object is seen through the instrument of a slave class and not of a proletariat.

In a slave-owning society the productive complex at whose kernel is the machine is still undeveloped. The machine is a mere tool, an outgrowth or auxiliary. Hence the supervision of the slave is not a matter of knowing the inner determinism of nature in a detailed fashion, but mainly a matter of conveying one's purpose to the slave, who fulfils it to the best of his ability. He is a human being; it is sufficient to give him a command. One sets

87

before him an aim. The organization of labour does not reflect as deeply as in bourgeois society the social division of labour. There are slave gangs—masses of men on whom the master's will is imposed with a lash—there is none of the elaborately differentiated organization of a factory staff springing from the necessities of the stuff handled and the machinery used.

Hence the movement of objective reality, Nature, can be satisfactorily expressed in terms of purposes, or ends. Fire rises because its destined place is above; heavy material falls for the same reason: it too seeks its 'appointed' sphere. The whole Universe is satisfactorily explained as a theatre of Will. Determinism or the interconnectedness of phenomena which is the most general category of objective reality in all societies, must in slave-owning society take the form of a pre-willed Fate. The Universe is a complete arrangement determined by some divine consciousness in a Universal Plan. In the same way accident is merely Divine necessity. As with Oedifices, accident is one will thwarting another; God interfering in the plans of man; or Moira in the plans of God. And because no market exists to cut man from Nature by a chasm and give the machine-complex an apparently self-moving power, the causality which is cosmic Will has to be perpetually *sustained*. The planets are urged on by spirits; the moving object perpetually needs force to overcome a resistance; there is a Prime Mover, God, who does not merely act as a universal co-ordinating force, but who actively pushes things on. This activity is not the laborious activity of a slave, directed physically upon an object, but the activity of will of a slave-owner, active upon nothing but the coerced mind of the slave. God, the

88

master, must always stand over slavish Nature, lash in hand.

Hence teleology is not opposed to mechanism, it is mechanism as it emerges in a slave-owning society, just as mechanism is teleology as it emerges in a bourgeois society. But the higher degree of interpenetration of Nature and Man which takes place in bourgeois society ensures that mechanism is a richer, more complex, more accurate picture of objective reality than teleology. However teleology contains more warm human qualities, it is less torn away from subjectivity, than mechanism; and even in bourgeois science it reappears in the spheres of change and higher quality.

Science in society is nature as it emerges in theory, but it can only emerge to theory in practice: it therefore rises through the producing class—the class that mingles actively with Nature. Hence the categories of science or 'things seen' *always reflect in a class society the particular conditions of functioning of the working class as seen by the ruling class.*

But the categories of mind, or 'things felt,' emerge directly from the consciousness of the ruling class. Just as it is the ruled class which wrestles with Nature, it is the ruling class which is conscious. Therefore the categories of mind—of philosophy, art, and mystical religion —*always reflect in a class society the particular conditions of functioning of the ruling class as felt by them.* Hence in a bourgeois society subjectivity is spontaneous and appears mysteriously containing its own inner sanction just as social desires appear spontaneously out of the free market. Its form shares the independence and irresponsibility as well as the ignorance of causation

89

which is the inevitable atmosphere of the capitalist producer.

Of course bourgeois production is in the first place centred round dead stuff, and does not to any large extent handle living matter in a fashion which would make categories of life important. Agriculture is not mechanized. However the later development of science leads to the study of biology. Thus biology keeps the categories of slave-owning teleology longer than physics.

Moreover teleology can reappear in biology in a specifically bourgeois form, which it would not be appropriate to discuss here as it would take us too far from our subject. All that need be said here is that bourgeois teleology, when applied to Nature, is by no means the opposite of mechanism, but reflects the categories of capitalist machine production in a different way, owing to the later stage of evolution of capitalism.

CHAPTER FIVE

THE COLLAPSE OF DETERMINISM

1. THE PROBLEM OF FREEWILL

IT is a remarkable feature of the present crisis in physics that it raises as central problems difficulties which have always been supposed to be the concern of philosophy. Against their inclinations scientists are driven to be philosophers: that is, they are driven to question the assumptions they had inherited unquestioningly from science; and now this has become a questioning of the very foundations of their world-view.

We have already dealt with the way in which the problem of the subject-object relation began to reveal itself in physics. And now another basic philosophical problem, that of determinism, is recognized by most physicists as requiring restatement in the light of new developments in physics.

If the development of macroscopic or relativity physics raised the whole subject-object problem anew, it was the progress of quantum or atomic physics which forced reconsideration of the problem of causality. Not only have both these problems yet to be solved within the limits of their special fields, but something like a Bohr's correspondence principle in philosophy is required to correlate the two fields.

In the nineteenth century it seemed as if the philosophical basis of determinism had been settled for science in the seventeenth century. Descartes, Locke, Leibniz, Malebranche, Spinoza, Hobbes and even Hume came substantially to the same agreement, in spite of the apparently wide differences between for example the materialism of Hobbes, the spiritualism of Leibniz, the scepticism of Hume, and the theology of Malebranche. Moreover in their philosophy they only gave a systematic basis to the empirical principles of Galileo and Bacon.

Physics developed on this apparently firm basis for three centuries, and it is only to-day that the whole theoretical basis appears to be shifting, at the same time as the foundations of bourgeois society itself are crumbling away. Here too then it must be that the categories implicit in bourgeois society are inadequate to the new content. The bourgeois world-view is becoming chaotic.

What exactly is it that is contradicting the old solution of the problem of determinism in Nature, and of its associated problem that of freewill in Man, and of yet another problem, often confused with the first, that of causality?

The concept of strict determinism which is at the root of bourgeois physics is most simply expressed by Laplace, who imagined a calculator provided with accurate figures of the precise velocity, mass, and position of every particle in the universe at a given moment. From this he could predict the whole future course of the Universe.

2. THE PRINCIPLE OF INDETERMINISM

This view has been undermined by Heisenberg's Principle of Uncertainty. This principle has proved of great value in experimental quantum physics. It states that the position and velocity of an electron or elementary particle can never be both exactly known. Only an approximate figure can be obtained if both are to be calculated, although either separately can be known to any required degree of precision. The more precisely the velocity or the position is measured, the less precisely the position or the velocity can be ascertained. The connection between them is the extremely small quantum of action—Planck's constant. This quantum connects position and velocity in such a way that the precise location of the electron, or the precise estimation of its velocity, involves a possible error in the other factor, of an extent determined by Planck's constant.

The importance of the principle is that it states an absolute or intrinsic uncertainty as a law of nature. This has been interpreted by many well-known physicists as meaning that indeterminism is a law of Nature.

The principle itself is the result of the development of the quantum theory, the basis of modern atomic research. This theory presumes a fundamental discontinuity in nature which hitherto had always been supposed to be continuous. The differential calculus and Cantor's definition of continuity are both irreconcilable with the quantum. It is now believed that all transactions between atoms are quantized—that is, that the 'action' involved must always be an exact quantum or integral multiple of quanta. Action is energy, or mass, multiplied by time. The quantum of action is excessively minute $(6.55 . 10^{-27}$ erg seconds) which is why this discontinuity in nature had not been observed before. The theory has received a body of experimental confirmation: Nature proceeds by jumps.

How do physicists account for the success of theories such as Newton's and Einstein's, which presume a basic continuity in phenomena and in practice seem to give an accurate picture of reality? This is accounted for by Bohr's Correspondence Principle, which states that in proportion as the number of atoms involved increase, quantum laws approach more nearly to the classical laws of Newton and Einstein. The sort of objects observed by classical physicists, such as earths and billiard balls, contain so many billions of atoms that the difference between quantum laws and classical laws is not measurable. The innumerable discontinuities overlap so to speak and become continuous. It will be noticed that this is only a probability. The discontinuities might coincide and be perceptible. But the odds against this are so enormous that the possibility can be neglected in the ordinary way.

Thus the old 'immutable' laws of classical physics are now held by physicists to be only statistical or 'probability' laws. They are only very likely to apply to mass phenomena, such as those of billiard balls and suns. In this they are like certain laws which had long been familiar to physicists, the laws of thermodynamics. According to these laws, heat and pressure in a gas are due to molecular movement. The molecules bounce and jostle each other like flying billiard balls. It is obvious that at any moment these billiard balls may all find themselves flying away from a surface simultaneously, and then there would be the 'miracle' of a gas without pressure. Gas pressure rises with an increase of heat because the molecules move faster.

Again, when hot and cold bodies are in contact, the faster moving (hot) molecules hit the slower moving (cold) molecules and speed up the slow molecules themselves slowing up as a result. However a number of collisions are likely to take place in which the slow molecule strikes the fast molecule in such a way that the fast molecule is still further speeded up and the slow molecule still further slowed. If by chance, in any one instance, all the collisions, or the greater number of them, were to be of this character, there would be the 'miracle' of a hot body gaining heat from a colder body. The nature of the circumstances however makes this unlikely. The more molecules, the greater the probability. This probability approaches certainty with ordinary objects; and so the scientist confidently predicts that the kettle of water will not turn to ice if placed on the fire.

However the classical laws of motion, such as Einstein's and Newton's, were supposed to be of a higher character than the laws of thermodynamics. It had always been supposed that the probability laws of thermodynamics could ultimately be reduced to certainty laws. Just as the insurance company's 'expectation of life' in the case of middle-aged men can be reduced to certainty in the case of a particular middle-aged man whom you happen to see being run over by a bus. For example, if we throw a die, we may say it is only a five to one *chance* which number turns up; but it would appear that if we knew the exact position of the centre of gravity of the die, and the minute irregularities of the surface of the die, the table, and the interior of the dice-box, and the exact path and velocity of the hand as it moved in the throw, and the mass and specific gravity of the die

95

and box, and the times involved, and the density and temperature of the air, and so forth, then we could, according to Hamilton's classical Law of Least Action, compute with absolute certainty what the number would be. In the same way it was felt that if we knew the life history of every molecule involved in a heat exchange we could estimate its behaviour and by summing the life history of all the molecules concerned arrive at an exact law, certain in its operation, which would however be so like the probability law in the case of visible objects that the certainty would not be worth the extra trouble.

Heisenberg's Principle of Uncertainty shatters this hope. The individual life history of the molecule depends on that of its constituent atoms, these in turn depend on that of its constituent electrons, and the life history of these according to Heisenberg's Principle can never be exactly known.

How did Heisenberg arrive at this conclusion? In this way: If a quantum of action is involved in all electronic transactions, is now securely established, any observation of a particle must involve the release or addition of a quantum of action to the particle observed. This will affect the particle correspondingly, like the recoil from a gun. Observation involves interference (the emission or reception of light, for example) and the quantum sets a minimum to this interference. The quantum of energy is a product of both position and velocity. The sharper we make the position the more we alter the velocity; the more exactly we observe the velocity the vaguer the position.

According to various physicists such as Jeans and Eddington, the conclusion to be drawn from this is that causality and determinism are no longer principles of physics, and it is possible to understand how the human will can be free.

3. THE SANCTION OF DETERMINISM

It is important to note that to the bourgeois determinism, causality, and free will have specific meanings which are by no means general to philosophy but suck their significance from the soil of bourgeois culture.

Determinism is to the bourgeois a characteristic of the world of Nature. It implies a necessary connection between events of such a character that the whole universe of events can be regarded as unrolled from the beginning according to inevitable laws. This predeterminism (as it in fact is) is symbolized by Laplace's calculator to whom the progress of the Universe can be exactly predicted once any section of it is known. He could of course with equal certainty move back into the past. Thus the whole of Being from beginning to end is necessarily determined by any one sector.

Evidently the doctrine of absolute determinism cannot be proved in practice, for to do so requires the unrolling of the whole of being. It is a principle. Nor can its sanction in the form stated be found in reason. How then did the seventeenth-century philosophers and scientists who laid the foundations of determinism justify their principle?

It was justified by an appeal to God's omnipotence and omniscience. Since God knew all that would come about, it was impossible that what would come about could be otherwise than as God foresaw them in His infinite reason. There was then a necessary connection between

97   H

events, which could not be otherwise. Hence natural laws were laws of God and also of reason.

The precise expression of this principle took various forms. With Malebranche and Descartes substance (matter) was so inert that it required creation anew for each moment of time. Conservation and change were the same. Hence Matter was from instant to instant suspended in God who therefore supplied a necessary connection between instants of existence.

To Descartes God was also the primary cause of motion: He put in a given quantity of motion, as well as of matter, and the Laws of Conservation of Motion thus expressed a Divine determinism as a lack of active interference by God in the Universe He had made.

Hobbes quite simply grounded determinism on the omniscience of God. God knew everything: hence everything was already settled in its minutest details: the details could not be otherwise than they would be.

Spinoza, in spite of the monism of his Universe, was also a 'strict' determinist. Contingency, efficiency, and freedom are to him only 'apparent'; they are aspects of the divine substance, which is completely determined. True the aspects of this substance are accidental. The existence of anything whose essence does not involve existence cannot be conceived as necessary, reasons Spinoza, and therefore must be accidental. None the less these things whose essence does not involve existence are treated by Spinoza as merely apparent; they are little better than illusions. The underlying basis of all phenomena is a substance, God, which is completely determined.

Leibniz, in spite of his idealist approach, equally bases

his system on absolute determinism. Although his monads are windowless, they appear to act and react on each other according to causal laws, because all has been arranged by God beforehand according to a pre-established harmony. This harmony is therefore an overriding necessity; it is absolute determinism. It is true that Leibniz attempts to introduce 'pure possibles' and a distinction between hypothetical necessity and absolute necessity. But the object of this seems to be as follows. At each stage the monad has before it various 'pure possibles' in the form of a choice of acts, of which it chooses one. Thus it is free. God however foresaw that it would choose this 'pure possible' and therefore there is a pre-established harmony. Obviously the only purpose of this qualification is to give a meaning to the conception of freedom and to prevent God from being Himself predetermined by the monads. It in no way interferes with the absolute predeterminism of the Universe.

Newton, although not an expert philosopher, accepted unquestioningly this method of approach. To him dead matter was inert, and all the transactions of matter were effected by spirit.

'We might add something concerning a certain most subtle spirit which pervades and lies hid in all gross bodies by the force and action of which spirit the particles of bodies attract one another at near distances, and cohere, if contiguous; and electric bodies operate to greater distances, as well repelling as attracting neighbouring corpuscles; and light is emitted, reflected, refracted, inflected, and heat, bodies, and all sensation is excited. . . .' (Gen. Schol., *Principia* III, p. 547.)

99

When the same kind of spirit is used to explain the force of gravity, it becomes the Divine Spirit—God. Thus in Newton's Universe too the particles are suspended in God, and derive the necessary connection of the events in which they participate from Him.

Hume is generally supposed to deny causality. Yet in spite of his scepticism he brings to his study of phenomena a naïve conviction of absolute determinism, and finds a justification for it in a principle of 'uniformity.' He denies causality; but bourgeois causality is in any case not the same as bourgeois determinism. The 'invincible uniformities' of Hume are therefore a subjective and individualistic form of strict determinism.

In the older theological form these uniformities had a sterner cast. Since the uniformity was a necessary connection mediated by God, one might look for laws of Nature, like those of Gravity, of a divine simplicity. In Hume's sceptical approach, there seemed no reason for such a faith; it is a mere individualistic foible. Hence Hume was the first positivist. With positivists, as with Cartesians or Newtonians, there is an initial presumption of absolute determinism in Nature. The former however make it a theological rule; the latter smuggle it in as a principle of economy or (in this case) uniformity. Hume is therefore a mechanist, but a confused one, as he imagines that by being sceptical of causality and freedom he is being sceptical of mechanism. But as we shall presently see, causality is not the same as determinism.

Kant carries this confusion of Hume's somewhat farther. Hume unconsciously sought for determinism and uniformity in natural phenomena and having found them believed them to be primary because of his mechanistic

100

bias. He might as easily have sought indeterminism and diversity and when he found them—as he could—insisted on accepting them as primary. It would have been equally valid. But Kant *consciously* seeks for 'causality' (i.e. determinism) in Nature; or rather he says that the mind necessarily imposes a deterministic scheme on phenomena. How the mind can do this unless the phenomena are of a character which makes this possible —i.e. already have necessary connections of some kind —is not discussed satisfactorily by Kant. Kant thus substitutes for a necessary connection between phenomena a necessity of the mind to see connections between phenomena. The mind takes the place of God in earlier philosophers. But the net effect is the same. All knowable phenomena—all that exist for us—are determined as absolutely as in the Cartesian scheme.

Berkeley adopts a similar view: all phenomena are 'caused' by a spiritual substance which is in fact God. Hence all events are grounded in God, just as they are by Newton or Malebranche, and this suspension of inert matter in God provides, because of God's omniscience, an absolutely deterministic framework. There is a necessary connection, *which could not be otherwise* than it is, between all events. It is true that from Newton to Berkeley there has been a change from corpuscles to phenomena (*esse est percipi*) as the basis of events, but this merely represents the divorce of the philosophy from experimental physics. Interest has swung from activity upon objects to enjoyment of objects. It is also true that from Hobbes to Kant there has been a progress from the omniscience of God to causality as a category of Mind. But in both cases an absolutely deterministic framework of events is sanctioned, and the change shown by Laplace, whose Divine Calculator was really only an exceptionally clever mathematician. God has already become a mere scaffolding for an hypothesis—the principle of determinism—and not a scaffolding for which there no longer seemed much need.

4. FROM GOD TO MAN

The eighteenth-century materialists adopted the seventeenth-century principle of absolute determinism. But they did not regard this principle as sanctioned by God; on the contrary it was a hypothesis about Nature, quite independent of the omniscience of God. Hypothesis is hardly the right word. Since they did not admit the possibility of its being modified by experience, it was a dogma. But it was a dogma in its own right. But it was not a dogma for which they advanced the sanction of an omnipotent God. Hence Lamettrie accused Descartes of bringing in God merely 'to please the priests,' to indicate that his system was consistent for God.

We have already mentioned the cleavage between the theological God of mechanism and the subjective God to whom a variety of personal characters is attached. In the age of the dogmatic materialists, the personal characters attached to God bore a close resemblance to those of Louis XIV and the *ancien régime* and it was for this reason more than any other that the encyclopaedists rejected God as a hypothesis for which they had no use.

Hence the dogmatic materialists are not so different from those seventeenth-century philosophers who give

their physics a spiritual tinge. All spare a belief in absolute determinism as universally applicable to matter (or
objective reality). Indeed this is equally true of Berkeley
and Kant. Hegel makes some distinction as to the spheres
of Nature in which mechanism is applicable; but fundamentally he accepts the same view. This dogmatic
mechanism is aside from the question whether they are
materialists or idealists. They are materialists if, like
d'Holbach and Diderot, they regard the stuff to which
these categories apply as being the sole objective reality;
they are idealists if they believe like Berkeley that the
stuff to which these categories apply is God. Or again
they may be Kantians and believe that objective reality
is unknowable in itself and that its deterministic categories are imposed on phenomena by the mind, or they
may be dualists like Descartes and believe that substance
is bound by God's determinism and yet that in some way
mind is free. They may be monists like Spinoza, and
believe that matter and mind are aspects of a Divine
substance who obeys his own laws; or monadologists
like Leibniz and believe that the particles are mindlike
and exist according to a pre-established harmony, matter
being only a confused perception of mind. They may be
positivists like Hume and see matter as a stream of
phenomena of which a principle of uniformity is the only
organizing factor. It is evident that the connecting link
in all these diverse philosophies is a belief in the validity
of absolute determinism as applied to natural phenomena.
It is therefore by no means the case that mechanism is the
distinguishing feature of dogmatic materialism as modern
philosophers hold. Russell for example suggests that
materialism as a philosophy is characterized by (a) The

sole reality of matter; (*b*) The universal reign of law. Point (*b*) in the light of point (*a*) means that the essence of dogmatic materialism is 'The universal reign of law in natural phenomena,' and we have seen that this belief was common to Descartes, Berkeley, and Kant, who are by no means materialists.

It is true that Lamettrie for example is a dogmatic materialist in his assertion of the universal reign of a particular type of law in phenomena, but Hobbes, in deducing determinism from the omnipotence of God, and Kant, in deducing it from the character of the mind which fits all phenomena in such a scheme, are equally dogmatic as to the universal reign of determinism in natural phenomena.

The precise way God or mind enters the scheme therefore is a question of the particular relation of God to social development. In the time of Newton, the bourgeoisie advanced against feudalism by backing a protestant God against a Catholic one. As Marx said, the case of Charles I, and the success of the Puritans, showed how divine inspiration from above could be countered by divine inspiration from below. Hence the God of this period is a God who expresses the expansive and practical relation of the bourgeoisie to Nature. He is a God who does not discourage experiment and speculation. It is precisely by experiment with Nature, by its command over the object, that the bourgeoisie at this period advances.

Half of religion is confused science—a muddled perception of objective reality. The perception of objective reality involved in bourgeois society is a deterministic one, and therefore at this stage the bourgeois confuses determinism or necessary connection with God. The theological God is simply the bourgeois name for abstract qualities of real matter, just as the personal God is a name for the abstract qualities of human society.

5. THE TRANSITION TO IDEALISM

Towards the latter half of the eighteenth century religion has become a reactionary force, and has allied itself with the landed aristocracy and finance capital. It is the enemy of the small bourgeois. Hence on the one hand we have the spokesman of the reactionary church, Berkeley, making God synonomous with matter itself, and not merely its necessary connection, and robbing science of its sanction in practice. On the other hand revolutionaries such as Voltaire and Lamettrie strip matter of as much of God as possible, and become Theists or atheists. With both schools matter is all which is not-mind, but with Berkeley not-mind is God, with Lamettrie not-mind is not-God, who, being neither matter nor mind, does not exist. Kant and Hume represent still other positions; with them the determinism is drawn out of the flux of experience. By Hume it is drawn out of this flux considered as outside the mind—from the far side of the flux, as it were. With Kant it is drawn from the mind's side of the flux. In both cases the mind of the individual replaces God.

Hence we have four possible derivaties of mechanism in an individual philosophy: (a) The nature of God—matter's behaviour explained in terms of God. (b) The nature of Nature. Matter's behaviour explained in terms of matter. (c) The nature of phenomena. Matter's behaviour explained in terms of personal experience.

(*d*) The nature of mind. Matter's behaviour explained in terms of the perceiving mind.

It is obvious that all these have as their foundation a dogmatic predisposition towards seeing mechanism in Nature. This is so fundamental, that it must be drawn from the very categories of bourgeois society. True there are differences in the whole trend of the philosophy. Newton's philosophy is one resolutely turned towards the object—Nature—and seeking to explore it—it is a world-grasping experimental philosophy. Berkeley and Kant are concerned primarily with the subject and with theory and therefore cut themselves off from the object and from experiment. The positivists go even farther in this direction, since in consistent positivism we can find no 'real' principle in phenomena, not even, as with Kant, a principle of the mind. We can only find principles of 'economy,' etc.

But this difference in the trends of these philosophies represents the difference between progressive and reactionary classes. As capitalism develops, theory becomes sundered from practice, subjectivity becomes the special province of philosophy and objectivity that of science. Local reactions within the general development appear as special solutions of the problem. All agree on the attribution of mechanism to Nature: the only quarrel is to how large a field is embraced by Nature, and this is a matter of how much is demanded by mind, which again is a reflection of how far theory has grown away from practice. It is only when bourgeois society as a whole is doomed, that the bourgeois categories of mechanism began to break down even for the special field of mechanism, the science of physics.

We have already explained how mechanism becomes the world-view of the bourgeoisie in regard to Nature; how it is no accident that capitalist society, which so developed the machine, sees Nature as a machine; and that when its theories of the machine are negated by capitalist crisis so its theories of Nature as a machine must at the same time also prove their contradictoriness. Strict determinism is a characteristic of bourgeois mechanism. It must be sharply distinguished from fatalism, which makes no distinction between Nature and man, whereas absolute determinism, by making determinism a characteristic of Nature, gives man an apparently divine power over Nature. Fatalism is appropriate to a class society based on open coercion of class by class, and hence is an ingredient in all pre-bourgeois religions; it is based on necessity conceived as will. Determinism is however based on necessity inherent in the object as contemplated; it is appropriate to a society in which coercion is veiled and is achieved through an administration only of objects, men being apparently entirely free. Man stands apart from the object and controls it because of his knowledge of its inherent necessity.

Strict determinism is not the sole characteristic of mechanism. With mechanism there necessarily goes a special distortion of the subject-object relation, in which the subject is torn from the object in a particular kind of way. We have already dealt fully with this.

The conception of strict determinism is also bound up with the problems of causality and freewill, and of probability, accident and necessity. These conceptions too are given in bourgeois society a special interpretation. They are all of a piece with the world-view of the bourgeoisie.

107

CHAPTER SIX

THE MEANING OF CAUSALITY

1. CAUSALITY NOT DETERMINISM

THERE is a tendency in modern science to use 'causality,' or 'principle of causality,' as equivalent to 'determinism.' As Eddington correctly points out, so far from being equivalent, they are incompatible. The relation of cause and effect involves a flow of power from the cause to the effect, and therefore a certain freedom on the part of the cause. But if every event is completely and necessarily determined, then how can any event be regarded as a cause, since it is absolutely determined from the start by prior events? It is not in that case the cause, but the cause is shifted back, and there is an infinite regress. Causality then as a universal principle equivalent to determinism has no connection with the ordinary relation of cause and effect except in the theological dogma that everything material is caused by God who is Himself His own cause. This was in fact the solution of Newton and other seventeenth-century thinkers. They did not, as generally believed, solve the problem of an infinite regress by an appeal to a cause historically prior. Such a solution would be unacceptable to a theologian. With them God was a cause logically prior—He was the necessary connection between events, which were therefore suspended in Him. In this sense causality has a different theological

108

meaning from determinism, although as far as science is concerned, the effects to be expected are the same—the universal reign of law. But when Planck and Einstein speak of causality, they use it in no theological sense, but as equivalent to determinism.

It might be argued that causality simply means the general affirmation of the principle 'Same cause, same effect.' But such a principle taken precisely is nonsense, for no cause (existent or event) can be exactly the same as another, otherwise how could one distinguish them? And if they are different, how can the principle be general or fundamental?

If the principle of causality is to mean anything, it might mean that everything in the Universe consisted of linked pairs of cause and effect. But each of these pairs would be self-contained and thus any pair would be unknowable to the others. Such a Universe would fall to pieces. And this in fact reveals the meaninglessness of any precise definition of the kind: 'same cause, same effect,' for in fact into every effect all the previous events of the Universe flow as a cause and lacking any one of them, the effect would be in some measures slightly different. This is in fact expressed in the General Principle of Relativity in which a fundamental local constant, the curvature of space-time, depends on the amount of matter in the Universe. Lacking one atom, events everywhere would be slightly different owing to their basis, the cosmic constant being different.

Hence 'causality' as a general principle based on 'same cause, same effect' either involves a contradiction or else it is a system of absolute determinism with the proviso that the system of determinism results from the omnipotence of God. 'Same cause, same effect' is then true only because it cannot be untrue, as there is only one cause, God, and one effect, the Universe. This is roughly Malebranche's position. In the latter case, determinism would be the general principle, covering the beliefs of the Cartesians, Hobbesians, and dogmatic materialists, whereas 'causality' would be restricted to that section of the determinists who give determinism a theological justification. Although it would be consistent, such an interpretation would not be helpful to those modern physicists who attack or defend causality. A third definition is possible, but it is a definition dependent on conceptions foreign to mechanism. Therefore until the foundations of determinism have been more deeply examined, I propose to use only the terms 'determinism' and 'causal relation' to express the first, a necessary connection between all events and the second, a relation between events involving a flow of power or efficacy from one to another.

A causal relation therefore has no particular relevance to the principle of causality, except that if causal relations exist, in reality, and the principle of strict determinism is also upheld, it must be shown how causal relations can exist within the framework of determinism.

The idea of cause and effect is not derived from pure cogitation, like that of being, it is derived from practice. I decide to move my hand, and it moves. I decide to shift an object, and it is shifted. In view of the importance we have already attached to practice in the subject-object relation, it is evident that this practical basis of the conception of causality is of significance. The idea of causality is not derived from the relations of objects between themselves, as is that of determinism, but from 'my' relation to objects. The causal relation in its purity already escapes from mechanism. For example, a pool of water is determined as to its outlines by the crevice in which it rests and the atmosphere which presses on it—ultimately that means it is determined by all that is not-pool. I can notice this in contemplation; it is a matter between the pool and the earth. (How I come to recognize perceptually the difference between pool and earth is not to be raised here, beyond stating that such a perceptual recognition must have a practical basis.) But in order to get a genuine feeling of causality, 'I' must will a definite . action and produce some corresponding commotion in objects—disturb the pool or break up the earth with a spade.

The first kind of determinism—between pool and earth, or pool and not-pool—is determinism within the field of vision, but excluding the 'I.' (For if you include the 'I,' part of the pool—colour, shape, and motion, is in the 'I,' but exactly 'how much' cannot be determined.) It is determinism wholly in terms of the object. It is identical in its laws to logical determinism, in which the Laws of contradiction and excluded Middle apply. However it is abstract, because it abstracts the 'I.' It is mechanistic. That is why the Law of Excluded Middle is only true in abstraction.

The second kind of relation is determinism which includes the 'I' and hence the whole of reality. But in it 'I' and pool flow into each other and interpenetrate. Hence the law of Excluded Middle no longer holds good. This is the dialectical determining relation, opposed to that of mechanism.

111

2. THE MEANING OF FREEDOM

This relation of cause and effect is closely linked with the idea of freedom, and this in turn is associated with efficacy or 'power.' If for example, without my volition, my hand rose to my shoulder (as it might in a nervous disease) I should not regard the 'I' as the cause. I should regard myself as compelled. Again if I was propelled by a shove in the back, I should not regard such a movement as caused by 'me.' On the contrary, I should regard my movement as caused by something outside my will. In both cases I should regard the 'I' as suffering and therefore figuring as an effect and not a cause. Hence if the causal relation is to have any meaning—i.e. any definition within my experience—it must mean the free action of a subject upon an object so as to cause a change in the object. It is impossible to give significant definitions of power and freedom which depart from this setting.

But evidently such a conception (if we are to accept it) is negated by the conception of strict determinism. For according to strict determinism, all particles of matter are bound in a necessary connection from the beginning of things. Hence the idea that one lump of matter, my body, can, as cause, freely produce an effect in another body, an object, is contrary to the principle of strict determinism. It is no solution to say that will or mind is not material, and therefore can be free of the treadmill of determinism. It is a poor sort of freedom, if I am bound hand and foot by strict necessity to assert that I can think what I like. And as I have previously mentioned, a genuine experience of the causal relation cannot arise in pure cogitation. It demands for its realization *practice*, or the production of change in matter—for matter is only a name for the category of objective reality. Hence the idea of freedom, which inheres in the causal subject-object relation, is negated by the principle of strict determinism.

3. SPIRIT AND SUBSTANCE

This was in fact realized by the seventeenth-century philosophers and none of them were able to give a satisfactory solution. They perfectly appreciated the difference between strict determinism and the causal relation, which involved freedom, efficacy and power.

We have already discussed the cleavage between theory and practice which leads to the separation of subjectivity and objectivity and in a class society. With the seventeenth-century thinkers, strict determinism or mechanism was a category of objectivity or matter, and rested on the theological God, who symbolized a still idealistic conception of matter. In the same way the causal relation—freedom or power—was given to subjectivity, and grounded on the personal God who symbolized a confused perception of humanity. But as we have seen, the causal relation is essentially an active subject-object relation—hence a contradiction was introduced from the start by this separation of the basis of freedom from the realm of necessity.

And in fact none of the seventeenth-century thinkers —or subsequent ones for that matter—were able to escape from the contradiction. Descartes tried the stratagem of supposing that God's determinism settled only the quantity of matter and motion in the world and that 'spirit'—the usual seventeenth-century name for subjectivity—could influence the direction of it. This was however contradicted by the Law of Conservation of Momentum and in fact Descartes himself admitted finally the weakness of the solution; and rested his faith in freedom on the fact that finite minds could not understand the contradiction but that God could be trusted not to deceive His creatures.

To Spinoza freedom can be a property of *natura naturata*, an aspect of *substance*, but not of natura naturans, the elemental and infinite substance itself. To put it in this way is however merely to soften the blow; freedom to Spinoza is only a seeming—determinism is Divine and real.

To Malebranche the idea that a finite existent could be a cause was a pagan belief. It was 'the most dangerous error of pagan philosophy.' In effect therefore Malebranche denies freedom, although he attempts to save it by distinguishing between occasional and genuine causes: this involves a denial of the causal relation in its fullness.

Malebranche regards the attribution of qualities or any form of causal power to bodies as 'pagan'—doubtless because it suggests something 'divine' and self-moving about them. Malebranche is correct as to the paganism of this belief. The more primitive the philosophy, the less the separation of subjectivity from objectivity. In primitive philosophies spirit is pneuma, breath—something tangible: matter is filled with 'occult qualities' and projected feeling—something living. Physics is the study of 'physics'—of living and growing stuff. The rigid separation made between inert substance and incorporeal spirit by a process of relentless abstraction is characteristic of bourgeois philosophy and another reflection of the cleavage of subject from object. Malebranche's dictum 'All bodies have no force to move themselves' is the basis of bourgeois physics. Carried to its extreme this meant that even to conserve itself, merely to exist, matter needed an exterior cause—God. In this respect seventeenth-century philosophy agreed with Plato—'existence is power.' But instead of drawing the conclusion that all that exists has power, they drew the contradictory conclusion—inevitable in the light of their epistemology, —that all that exists is maintained in existence from moment to moment by the power of God.

Leibniz's solution is ingenious. Since all the monads are sentient mind-stuff, it would appear that the problem of dualism does not arise. But with seventeenth-century thinkers matter is not so much not-mind as mechanism, or necessity—the theological God. Hence Leibniz is unable to help introducing objective reality into his mentalistic Universe under the guise of God, to whom all monads turn their windows. God with Leibniz plays the part of matter. It is true that the monads do not affect each other causally, but that merely shows that Leibniz has accurately grasped the fact that objective determinism does not include the causal relation, but excludes it.

The monads may seem to be free, because they unroll their history from within themselves. 'Pure possibles' exist at each stage, and though they have their reasons, each monad inclines to choice under no compulsion of necessity. In fact however this is not a causal relation: for while a causal relation is not in its purity a wholly objective relation, neither is it a wholly subjective one.

115

The monads however are prisoners of their pasts. The causal relation involves activity of a subject on an object; but the monads are windowless, except to God. Of course in reality subjects can only affect each other through matter and since Leibniz's God is a name for matter, his Universe is correct. But it is turned inside out. And in any case, whether 'pure possibles' have any meaning or not, all has been foreseen before by God, whose contact with all monads ensures the dove-tailing of actions and the pre-established harmony. Determinism overrides all. Thus in spite of its wild romantic air, the monadology of Leibniz is simply dogmatic materialism under another name, and with 'God' substituted for the premise 'matter.'

To Newton matter was a dead inert substance, quite different from the live matter of the Epicureans. Causal relations were therefore the work of 'most subtle spirit' in everything—but since this spirit also was determined in its actions, it was either itself inert, and determined by God, or else was God. Newton seemed to incline to the latter decision; and in either case it will be clear that he denies the causal relation apart from God.

Hume faced with the choice of all bourgeois thinkers —determinism or the causal relation—also chooses determinism. In his system it appears as uniformity or regularity of sequence. This by no means involves a causal relation, which rests on the idea of power. The sunrise may follow the crowing of the cock but it is not a causal relation. If however cockcrow and sunrise is an 'invincible' uniformity, it is a deterministic relation because it is necessary.

Of course it could be argued that only causal relations

116

generate invincible uniformities, but that would be a principle and not a deduction.

If Hume notices a uniformity in events he must also notice a causal relation between himself and objects. To a sceptic both should have equal validity.

Forced to choose between the two by his world-view, which cannot admit both without contradiction, he chooses determinism and rejects the causal relation. Thus his scepticism about the causal relation is the result of a pious faith in determinism quite as pure as that of Malebranche.

4. THE FLUX OF PHENOMENA

However there is a difference which entitles us to regard Hume as marking a new stage in the unfolding of the bourgeois contradiction. Earlier thinkers based their belief on the invincible uniformity of Nature—i.e. in determinism—on a belief in God. This meant that they believed that the uniformities were not just uniformities, but that 'behind them' was an objective law. To a Humean there can be no real difference to be sought (except as a matter of prejudice) between a uniformity like that of cockcrow and sunrise and that of thunder and lightning: but to a Cartesian or Newtonian there can be an 'underground connection' and this determines whether the two events are 'really' connected. It is the importance of this kind of underground connection which people have in mind when they feel a principle of caus- ality to be different from a principle of determinism (or invincible uniformity). There is no 'reason' why the latter should be reducible to any law, much less a compact or beautiful law—yet this 'faith' always urges forward the scientist. Physics is a continuous drawing forth and codifying of these underground connections between phenomena. Hence a belief in determinism coupled with a lively belief in the possibility of discovering connections of t'. is kind, does entitle one to distinguish between determinism and causality. The former might merely mean that one believed in invincible uniformities; the latter means that all uniformities can be reduced to compact laws, or have underground connections. In seventeenth-century philosophy these underground connections are furnished by the theological God. It is simpler and less confusing to suppose they are furnished by matter, or objective reality, whose characteristics are described by the laws of connections.

We do not then get left over an inert substance, parent of Kant's 'thing-in-itself.' Such a belief in causality is not therefore necessarily grounded on theological faith: it may be equally grounded on materialistic faith. Lamettrie and Diderot equally believed in underground connections, but in their case these connections are provided by matter, considered as being a cause in itself, and not like Newtonian matter, quite inert, or Cartesian substance, merely extended. This belief in a prior regulating reality back of the stream of phenomena, giving them an underground connection, is lively in the dogmatic materialists. In fact there is no real difference between the matter of the dogmatic materialists and the God of the Newtonians and Cartesians except that there exists, in addition to God, with Descartes a quality called extension, and with Newton an additional quality, mass.

These were later to cause trouble to physicists and hence the dogmatic materialist's faith was superior to

118

the theological mechanics on the score of methodology. But better than either was to ground the interconnectedness or basic commonness of phenomena, not on a faith in either God or matter, but on activity on matter—by exhibiting in practice the transmutability of stuff according to certain strict laws.

Thus the 'theological' God of the seventeenth-century thinkers is really matter: but of course it became much more barbarous and unscientific when it was equated with the personal God of the myths and mystics. However the practical bourgeois physicists stood no nonsense from God in their particular domain—He was ruthlessly stripped of mercy, love, an only-begotten son—everything but His material determinism. He was a scaffolding.

Thus both dogmatic materialism and dogmatic Cartesianism have a 'reason' for looking for compact and 'beautiful' underground connections between phenomena—the fundamental faith of a scientist. The positivist has no such reason. If nothing exists but phenomena, there is no cause why one should happen rather than another; why uniformities should exist, why they should be expressible economically, why any uniformities or relations should persist; why one should be able to orientate oneself in the Universe at all. Here all physical research is mere prejudice—its successes are mere lucky accidents. Anyone who knows the profound faith of the physicists in the existence of such laws, will realize how reluctant he is to accept such positivism. The experimental physicists such as Planck (as contrasted with Mach, a pure theoretician) reject such world-views, and if they are driven by the breakdown of bourgeois categories to accept them, then, to justify their prejudices, they again call in God to sanction a belief in uniformity. But now he is a God *from the other side*. Because, by the development of positivism, phenomena have become a screen, at first holding the unknowable thing in itself and then *nothing*, the God whose existence assures the physicist of a worthwhile end to his labours is no longer the 'most subtle spirit' of Newton, 'the spiritual substance' of Berkeley, the natura naturans of Spinoza, or the stern impartial Necessity of other philosophers, he is a Mind, remarkably like that of the physicist. For example with Jeans he is a mathematician, and this assures Jeans that he will find equations everywhere in phenomena: thus Jeans is saved from the positivistic nightmare. To Russell and Eddington there is behind phenomena 'mind-stuff' resembling one's own consciousness. But all such introjections of one's own mind behind phenomena to take the place of deleted matter (which is what all these theories amount to) represent a certain falling-off and disorientation as compared to the earlier physicist's robust viewpoint. They were searching for something prior to and more complex—and yet also simple and more necessary—than consciousness, whether they called it matter, or God, or reality. It was a prospect to expand the mind and lure it into practice. Modern scientists however by such a creed are merely searching for something already in consciousness, even if it is only implicit. Hence a strong tendency, clearly shown by Eddington, to extract truth by mathematical manipulation at the expense of experiment.

Thus the mentalism of Eddington is different even from the mentalism of Berkeley or of Leibniz. Berkeley and Leibniz both believe that lying behind phenomena, and necessarily determining them, is an objective reality broader, prior and simpler than mind, although discoverable by it. With Berkeley this reality is spiritual substance; with Leibniz it is the God to whom all windows open. But with Jeans and Eddington and Russell objective reality is indistinguishable from mind: hence it seems rather a waste of time to engage in practice and in experiment—to go out into phenomena to discover what is already in essence in the subject. Evidently this philosophy is a symptom of a tendency for theory to drift away from practice in physics—the 'descientification' of science. Theory remains attached to practice on more and more limited and specialized fronts. If such an attitude were to become general in physics—and there is every sign that it may—it would be serious for the whole future of science. It might lead to a withdrawal of science from experiment into a barren theorizing, and this itself would be the result of a general reduction in experimental effort, a slackening of the magnificent tempo of research characteristic of the last century. Economic conditions are already beginning to start such a movement in capitalist Europe. Money spent on research is diminishing. Thousands of students each year are denied by unemployment the opportunities of experiment. Are we witnessing something like the scholasticism of science— a treason, not of the clerks but of the investigators?

5. THE MOVEMENT OF LOGIC

Kant's critical idealism in fact cleared the ground for such a development. With Kant there is the flow of phenomena but they flow in a framework of determinism:

121

this however is an imposed framework. However the very fact that phenomena are susceptible of such a framework makes them lawful; and the fact that the mind must impose this framework is yet another law of reality. Hence necessity is still grounded on 'faith' in the existence of physical laws, and this faith is reflected in the queer appendage which exists the other side of the screen—matter 'on trust,' or the unknowable thing in itself. This thing-in-itself is a kind of pledge for the honesty of the law of determinism. It is a reflection of the mind the other side of the screen—the two necessities supplement each other.

It remained for Hegel to point out the non-existence of the unknowable and delete the thing in itself. This did not however lead to positivism, for he substituted for the necessity of objective reality or mechanism the necessity of subjective reality, or logic. Phenomena unfolded themselves with the determinism of logic. But in doing so, the mind dissolved into Ideas which began to lead an existence independent of the subject. They were absolute Ideas and unfolded themselves according to their own necessity. They had therefore become objective reality, and Logic had become equivalent to the God of Malebranche, the substance of Spinoza, the spirit of Newton and the matter of Diderot. There were two important differences. These Ideas, just because they had sprung from the loins of the human subject, changed—they unfolded themselves. Cartesian and Newtonian objective reality, being grounded on God, had always been eternal and changeless. Thus objectivity for the first time had been given the quality of self-evolution and change, because of its former history attached to the subject. It had become dialectic—or, as Malebranche would put it—pagan. It had returned to the live matter of Epicurus, but containing within itself all the subjective complexity developed in the interim. Object is the only form in which the bourgeois can know it: that of mechanism. The object can only blossom into consciousness of society again in the form in which the proletariat can know it, as the Nature upon which Man is active in the concrete relation of society. But during its development in the human head, it had torn itself away from any outside reality on which to 'practice.' Thus it lacked the very means of self-discovery and self-development which is afforded by practice on objective reality. The Hegelian Universe could only unfold with beautiful accuracy what was already in it and then stop—with Hegel. It could not drag into itself any fresh knowledge from outside (as man does by practice on objects) because there was no 'outside.' Hence Hegelianism could not be a physicist's creed, for it denied the need for physics. It could only be a speculator's creed.

Yet its criticism of the Kantian thing-in-itself was bound to leave its mark on physics. It had pulled aside the screen of phenomena behind which it lurked, and had shown there was nothing there. Matter ceased to exist for the philosophizing physicist, and yet he could not believe that phenomena were merely the unrolling of absolute ideas for then there would be no need for physics. Hence he had to accept a faith in 'underground connections' between phenomena grounded on the supposition that phenomena were the aspects of some fundamental substance like consciousness.

It is true that with Hegel, who completed the contra-

123

diction between subjectivity and objectivity, it was possible to pass beyond, to the Active subject-object relation of dialectical materialism. But to do so is to pass beyond the categories of bourgeois physics, and shake off the limitations of mechanism. It was impossible for the bourgeois physicist to do this, for he would have ceased to be bourgeois and become revolutionary in his relation to the whole of bourgeois ideology; and to do so would first have had to become revolutionary in relation to real society. It was only possible to pass from Hegelianism to positivism. Just as Hegelian dialectic, in which objective reality for the first time acquires of its essence the power of self-development, reflected the rapid evolutionary period of capitalist production, so the later stages of positivism are coincident with its decay. The bourgeois ceases to have a dominating world-view because he has ceased to dominate nature through the channels of society: he has ceased to control the machine in the old way—the machine controls him. His grip on the object has slackened. The flux of phenomena in positivism reflects the flux of events as seen through the meshes of a collapsing economy. The bourgeois world-view has lost its strong objective organizer, matter, because in real life the bourgeoisie has ceased to remain in close contact with the object, which is slipping wholly into the organization of the exploited class. The bourgeois remains outside, unattached, and this is expressed in a similar idealistic detachment from reality of his theory, or seen world. The deletion of strict determinism from the bourgeois world is the removal from his consciousness of the object in the only form in which the bourgeois can know it: that of mechanism. The object

124

can only blossom into consciousness of society again in the form in which the proletariat can know it, as the Nature upon which Man is active in the concrete relation of society.

It was said earlier that although determinism and causality were the same, if interpreted strictly, and in either case exclude a finite causal relation, still it was felt that events might be connected deterministically such as for example cockcrow and sunset and thunder and lightning.[1] When physicists talk about strict causality they generally have in mind a belief that all relations can be reduced to the second type. It will be interesting to see if there is in fact any difference between the two kinds of relations.

It has already been said that our most genuine experience of the causal relation is when we freely will to produce an effect and do so. This 'genuine' causal relation involves a feeling of freedom and power and we saw that if this were lacking the causal relation would be reversed and we should feel the passively effected object of an exterior cause. And yet this feeling of freedom seems contradicted by an invincible uniformity in outer reality.

Let us examine this causal relation more closely. The essence of it is that the subject, by its activity, produces an effect in the object that was not previously present. There is an ingression of novelty. This novelty does not belong to the subject, because it is an effect 'on' the object. It does not belong wholly to the object because it is 'produced' by the subject. It is therefore a joint product of the relation. The effect is therefore a novelty

[1] Unrevised Section starts here.

125

which emerges from the free activity of the subject on the object. This constitutes the essence of the causal relation.

Now this is evidently different from the determining relation of necessary connection. This is a pure relation of negation. The pool is determined by all that is not pool, and the crevice in which it lies by all that is not crevice. The relation is necessary, for no other mode of determinism is imaginable. And yet we cannot say strictly that the crevice causes the pool or *vice versa*. It is a purely logical relation, a determinism by negation. and evidently it has the necessity of logic. In any given field of consciousness, any object is determined by its negation in this way.

But suppose a wind ripples the surface of the pool. Then we say that the wind is the 'cause' of the ripples. For a change has taken place; a novelty has emerged which is a relation between pool and wind. Hence wherever we see the emergence of novelty, we 'introject' a causal relation like that we have ourselves experienced. When the physicist talks about causality he has therefore a relation of this character in mind, and imagines it as universally obtaining between events.

What right have we to introject in this manner, and endow a part of objectivity with the qualities of a subject in relation to another part of objectivity? We can only do so as a result of practice. If for example we make ripples on the pool with a stick, and produce an effect, and feel the wind on our cheek, and sustain pressure, and then press ourselves with a stick, we imagine the wind acting in our place as a cause upon the surface of the pool as object.

Examination shows us that in fact our whole field of perception is made up of practice or the results of practice on the object. If, for example, we see a pool lying in a crevice, we only do so because we have in the past causally explored the surfaces of object, water, and the interior of crevices. Thus we build up the qualities in the field of perception by memories of causal relations with outer reality. Most of a baby's early life is spent in building up its field of consciousness in this way. Hence all the qualities of the seen world are products of causal relations with the object, and thus even our determinism springs from this. A quality can only be determined by all other qualities not it, and this recognition and distinction can only be the result of causal relations with outer reality. Some of them are produced by us as cause; others are produced on us as effect—as, for example, when a new colour enters a child's field of consciousness for the first time. The changes we ourselves produce are of special importance, as we tend to make them the framework of reality.

Hence determinism, although it is precisely the same as strict causality in theory, and excludes the causal relation, can be given a meaning as soon as we study its generation in theory by practice. Determinism is merely the logical characteristic by which we denote existents in the field of consciousness. Since in a static field of consciousness an existent is absolutely and necessarily determined by the remainder of the field (p or not-p) the parts of objectivity appear to be necessarily connected. This is the source of the conviction of strict determinism. It is a mere character of conscious reflections of objectivity.

127

But because it is purely a logical form, it is without content, and is of no value to physics. 'Whatsoever is, when it is, is necessarily so as it is' is in fact an old scholastic dictum. Anyone can understand that all that is p is determined by all that is not-p, and that if everything not-black is sorted from the Universe, it leaves only black. Thus determinism in its strictest form is nothing but a law of thought, the statement in physical terms of the principles of Exclusion and Contradiction.

For this very reason, however, strict determinism is not an adequate basis for physics because a logical law operates entirely within the realm of theory, and merely soils the 'premises,' i.e. what is already in the conscious field. But physics is concerned with the 'cause' of the conscious field; that is, aware that changes are produced in the conscious field (objects move, etc.), aware that it is the subject of an effect, it asks what is the cause? And it can only answer by practice, by itself being driven out to fill the role of cause, and itself produce changes in its own conscious field—i.e. changes in 'matter' in the source of changes in phenomena. Hence there is a difference between the principle of determinism and causality. The latter asserts that there is an underground relation, or connection, between changes in its conscious field (among phenomena) similar to the causal relation 'I' have experienced as subject. Hence to the logical principle of determinism, is added a general declaration as to the existence of connections or causal relations between phenomena which is in fact an assertion of the existence of matter. This assertion with most seventeenth-century physicists took the form of the assertion of the existence of all phenomena in God. With the dogmatic materialists

128

it took the form of the assertion of the existence of all phenomena in matter. There was no real difference between the two for both were mere means, mere principles asserting the general presence in nature of causal relations of a subject-object character. They had to be filled with content, and this could only be done by experiment and hypothesis. These two are a general instance of the subject-object relation. An hypothesis lays down a certain world-view—an experiment either confirms it or contradicts it. In the latter case the hypothesis must be changed to follow practice.

A hypothesis makes an abstract world-view—this is one that deals only with a certain sphere of qualities. If it were to deal with all (apart from the unhandiness) every event would contradict it because it would be unincluded in the view. But the abstract world-view has holes and all events taking place in the holes demand no change in the hypothesis. A pleasure in an unexpected place cannot upset a physical hypothesis, but a star in the wrong place can.

Thus the conception of causality which involves the existence of objective reality (otherwise there is no causal relation) has after all a meaning distinct from determinism. It has no meaning for theory; but it has a meaning for positive science; theory plus practice. And we saw why it had at first a theological basis. Man required an assurance for the production of activity in matter, because by letting theory grow apart from practice, he had robbed matter of life.

Causality as the framework of science has therefore a practical significance. It is simply matter, the thing-in-itself becoming the thing for us. The unknowable thing-in-itself cannot exist, for even to know it is unknowable is to know something about it. But we know more. We know that treated in certain ways it reveals certain qualities. We change it—produce qualities—ripples in water, synthetic dyes, artificial rocks, and sun images. The particular causal relations involved in these productions of qualities, generalized and systematized, therefore give quite a lot of information about the unknowable thing in itself.

Hence 'naïve realism' or materialism is justified not by theoretical arguments but by practice. By continually 'changing Nature,' by continually producing effects and phenomena we learn the qualities of matter. Matter is a mere name—as vacuous as not-matter. It is only matter in its causality, in the relations which make qualities appear, that becomes rigid and fleshy and really existent. In this sense existence is power.

CHAPTER SEVEN

*Bell and earth—Universal determinism . . . contra-
dictory to causality?—Like and unlike—Time and
Space—Mutually determining—Modern views:
Probability (accident)—Determinism—Causality—
Freedom—Iron bourgeois determinism—Freedom:
ignorance of necessity*

WE start speculation with quite a healthy belief in the
existence of matter because we have been changing
Nature, producing phenomena, since babyhood. The
habit is already ingrained in us. Science merely refines
and increases this knowledge of matter by putting at our
service the codified results of generations of social
experience in changing Nature, and also still more
complex and penetrating apparatus for changing Nature.

Because our first causal experiences orientate our
whole world-view, we think of causality basically in
effortful kinetic terms. Force, energy, activity, power,
become basic categories of scientific thought. Why?
Precisely because they are associated with the product of
novelty. Hence, seeing novelties emerging in the world,
we explain them by causal relations between existents and
imagine them produced in dead matter by force, or
energy. Even in quantum physics activity, in the form of
action, is a fundamental conception.

This conception of activity, introjected as a result of
the subject-object experience of the causal relation, is
fundamental to physics, because it marks the difference

131

between a logical determinism and a scientific causality. It determines the whole way in which science is orientated.

For example, we hear a sound. The sound is traced by physics to a chain of cause and effect which links the excitations of the auditory nerve with compression waves in the air, these in turn with percussion waves in the bell shirt, and this in turn to the relative motion of shirt and clapper; again this is traced to the person swinging the bell. This is a typical scientific chain of 'causation,' of the kind which science would be reluctant to give up, and it is obvious that it is dominated by the concept of activity. Its abstractions and symbolizations (and the chain is highly abstract and symbolic) only acquire a meaning from the concept of activity and the connection of this particular event with various scientific experiments, each embodying a causal relation, in which man has exerted an activity on nature. Man has or can produce compression and percussion waves, excite auditory nerves and swing bells and hence the simple quality sound becomes endowed with a symbolic musculative which fills the chain with rich qualities of activity in spite of their apparent aridity. Hence *in practice* causality is seen to be a very different conception from that of determinism; it is determinism full of a history of practically experienced causal relations.

If however we adopt the purely deterministic approach it is plain that the causal chain we have outlined is incomplete and misleading. For the air and bell and ear being where it is depend on the location and movement of the earth, and on all the universal being disposed in exactly the positions they are. If we are to take the principle of relativity at all seriously, any change in one instantaneous point would mean nothing else could be precisely as it is. Moreover historically the ear and the bell and the air are the product of a long process of cosmogony which at each stage determines the following stage. Hence logically, the causal chain of ear-air-bell is a pure and arbitary particularism. Everything in the Universe is a 'cause,' i.e. a determining factor in this sense, not merely now, but in the whole past. What the bell has been, even, determines what it is now. And of course this is a mere restatement of the logical position, all that is not ear-air-bell determines what is.

At first sight there seems a certain richness in this logical determinism as compared to the ultimate mathematical bareness of scientific causality, which strips the existents concerned of quality and reduces them to equations. The purely logical determinism takes all as all, the world in its fullness—all qualities and all events. But precisely because it takes them as one *static* mass, the qualities vanish like a blown-out candle-flame. For we saw that even to distinguish pool from crevice as separate qualities, involved at some time a causal experience of pushiness and crevices and hence the generosity of determinism in taking the whole sum of qualities amounts to just this, the division of the world into p and not-p—and even to do this involves some experience of the causal relation—some active and appetitive struggle of Man with Nature which enables him to distinguish p from not-p.

But the ear-air-bell selection, because it is impregnated with activity, does stand out as an organized 'whole' from the background of the Universe. Something is 'happening' against a background of not-happening.

133

Of course there are happenings in the background but they are dismissed because they do not participate in the same sphere of qualities. They are light green flashes when we are concerned with the production of red. Hence as regards this sphere of quality in which we are interested, the ear-air-bell domain, the background is static. Thus it is that causality comes to impose itself upon a purely logical determinism and a 'restriction' of quality comes to involve a generation of quality. If quality is unrestricted, we get the colourless universe of p and not-p. If it is restricted, as part of the generation of a system of causal relations, it enriches the Universe with qualities, because enrichment can only come from experience, and the causal relation is experiential. Knowledge is knowledge through objects. The causal relation and experience both demand as one term the object. And the relation so far as it can be separated from the terms, is just this—change, a novelty. Hence science interprets the world by changing it.

To detach the causal plexus of ear-air-bell completely from the background is a fallacy. It is distinguished from the background, and yet arranged along its grain. It is determined by a (relatively) static background but is itself the theatre of an inner *activity*, of the subject-object kind. The activity is indicated by the glow of a new quality, which it has produced from its centre. As between background and complex there is an opposition, and also a mutual determination, in other words, they constitute a unity of opposites. This unity of opposites is however purely logical and merely the subject of determinism. Inside the complex itself however there is a supra-logical relation (using logic as formal logic).

This relation is exactly what is meant by causality and is of interest to science. It is inner activity and the production of a new quality against the relatively unchanging background. ('Every action has an equal and opposite reaction.') This expresses duly and formally the formal movement behind every quality which demands from science inclusion in physical causality, over and above its inclusion in logical determinism. In future, instead of calling such a relation supra-logical, we shall call it dialectic, because it is of a kind which is the special feature of Hegelian logic. It may seem novel to suggest that physics is concerned with the production of novelty, as hitherto it has been supposed that its role has been to strip phenomena of every quality (first taste, scent, colour, then shape, mass, velocity) and reduce it finally to 'mere' equations. But we shall show that 'mere' equations differ from logic precisely in this, that they are designed to express the production of novelty in the most general way.

Consideration will show that as soon as we leave the field of logic, and enter that of practice, we find that determinism, as a strict definition, is self-contradictory. For example if a series $A B C D$ . . . etc., is generated according to the categories of determinism there is a necessary connection between them of such a character that given A, then B, C, D, . . . etc., must necessarily arise. In other words the qualities of B, C, D, etc., must already exist in A. And given B, A must have necessarily been its cause. And given C, B and A must necessarily be contained in it.

What then is the difference between A and B? A contains its forebears and B, C, D. B contains A's fore-

135

bears and *A* and *C, D.* . . . Hence the difference is this, that *B* contains everything in *A*. How then is it distinguishable? Yet it must be distinguishable in some way for if there is absolutely no difference they are the same, and the series *A A A A* has no meaning. It therefore contains something not-*A*, which as it were constitutes its *B*-ness, and there cannot be a necessary connection between *A* and *B* of a character to produce this *B*-ness. There is a production of something new. However there is also a necessary connection, the *A* in *B*, which enables us to call them a series.

For example in the series 1, 2, 3, 4 unity is the necessary connection between 1 and 2, and 2 between 2 and 3, and 3 between 3 and 4. However 1 is the necessary connection between 1, 2, 3 and 4 and 2 between 2, 3, and 4. In other words a vein or persistence of 'like' quality is a necessary connection between existents. In the same series the second unity is the novelty and in the 1 2 series, and the third unity in the 2 3 series. Thus an element of unlikeness or novelty forms part of the necessary connection and yet is wholly different, in fact opposed to it.

But part of the like in the 2 3 relation was unlike in the 1 2 relation. In other words, in any series an unlikeness, after emerging, is carried forward as a like into future events. Hence if we travel back down a series we find likes continually splitting into unlikes and at each stage an unlike disappearing and travelling forward, unlikes continually emerging and gathering themselves into likes. Thus we see that in spite of its apparently formal characteristics, the series of integers are in fact the theatre of a continual activity, unfolding new

136

qualities from its relations. It is however also the vehicle of a strict determinism that of unity (or the fact that it is a series of *integers*). If all integers were thrown on the table, there would be only one way in which they could be arranged.

> *Distinction between determinism (already there) and novelty not there—Yet also prediction—Other forms of determinism all require a basic figure not itself apprehensible: empty of quality: leaps into new sphere—Hierarchical or spheres—10's or* a *and* b *or so on—Number very different to logistic causality and determinism—Time and space*

But it is evident that in order to do this, all the integers must be already there, with their qualities known. But if they are not there, they can only be unrolled from the early members of the series in a predictive way. For example by the formula $+ 1 \ldots$ However this formula merely contains the addition of like, and hence although the series can be formally unrolled in its fullness by the addition of 1 2 plus, each number cannot be known in its individuality until it actually arrives. For example until we have arrived at 2 or 3 or 79 their peculiar individual qualities (oddness, indivisibility, factor, etc.) cannot be genuinely known.

Thus the 'necessary connection' of determinism in the series of integers only covers the unrolling of like qualities. These can be fully predicted—i.e. there is a unique and necessary determinism. But each number also has a uniqueness, a novelty, which can only be known when it emerges. As soon as it emerges however it ceases to be 'unlike' or 'new' and is ingathered into the system of

137

determinism. Thus as far as the past is concerned all qualities of whatsoever kind are ingathered into a solid crystal of determinism, but as regards the future, we can only predict a shuffling of the old qualities with the certain knowledge that each shuffling will be the theatre of an inner activity unfolding unknown qualities. As soon as a quality is gathered into the past (or crystal of like) it becomes quantity, but quantity is continually generating quality (the new, the activity). That is what time is, and that is the meaning to be attributed to Time's irreversible arrow. This is what the distinction between past and future is.

It is for this reason that if we attempt to achieve 'pure' quantity and strip being of quality, nothing is left except numbers and the series of integers is itself a pure dance of quality. For every quantity emerges historically according to the series as a quality, and hence to strip existence of all quality leaves nothing but a pure being indistinguishable from not-being.

The bourgeois conception of determinism, owing to its formally logical structure, is however unable to attach a reality to Time and evolution. Hence it always imagines the events 'lying on the table' like the series of integers, and noting the unique necessary connection of the series, it assumes that the future can be predetermined at any stage from the past. But to do this in its entirety is to suppose that the new quality in events is already known —i.e. that contrary to our definition, the future has become the past. But to say that we can predict the future when it has become the past is absurd.

However it is possible to unroll the future from the past without self-contradiction in the limited sense of

138

unrolling only the necessary connection—that is, the like. In other words prediction is purely quantitative. Of course this quantity includes qualities already known, which are gathered into a consistent mutually determining web—as expressed by the series of integers. Is it possible to do this even completely? By bourgeois criteria it is, but we shall presently show that this is a contradictory conception.

The series of integers is disconnected. But this is an inevitable ingredient of their necessary connection. We saw that for 1 and 2 to be necessarily (i.e. uniquely) connected, 2 must consist of 1 and not-1, but 1 does not consist of 2 and not-2. Thus the determinism is unique. There is ostensively no such relation between A and B and hence there is no inner relationship; it is a purely formal one. B does not consist of A and not-A, or *vice versa*. Yet they have likeness—both are letters, there are similarities in shape, they may be contiguous. Make these differences ever so small, until they are the same letter; there is no unique connection between them until they become the same letter, and then you cannot speak of a connection between an object and itself, except as it unrolls itself in time. But if we postulate no difference at all, there is no difference even in Time. Hence there is no necessary connection.

Unique necessary connection or determinism therefore does not inhere in relations of the A-B character, but of the 1-2 character, in which there is a kind of one-way Chinese boxing (1 in 2, 2 and 1 in 3 and so on). But this one-way Chinese boxing is precisely Time's arrow. The essence of the unique necessary connection is that 1 should be negated by not-1. (A new quality emerging.)

139

For if it was merely 1 they would be the same event; there would be no means of distinguishing them and we have already seen it is impossible to talk of a necessary connection between an object and itself apart from Time. But since at this stage of the development of the series nothing but 1 is known, the not-1 must also be 1, yet the result is not 1, but 2.

The essence of the movement is that 2 includes a not-1, an element of difference. Without that it is indistinguishable and hence there is no necessary connection. The 'farther apart' are the events (i.e. the greater the element of not-1) the more apparent the necessary connection (1). (e.g. $1 + .. 99$). The nearer they approximate, the less apparent the necessary connection—e.g. 1 and 2 the less apparent the necessary connection. Hence discontinuity is an essential part of determinism and when it vanishes determinism vanishes and the object merely becomes itself. It is true that physics has hitherto based itself on the assumption that the greater the continuity (the approach to a limit of the calculus) the stricter the determinism. It has been supposed that determinism depends entirely on the principle of continuity in phenomena. Our analysis shows however that the opposite is the case, and determinism depends on discontinuity (real difference or novelty). To deny necessary connection is to deny the emergence of novelty and the reality of change. This in fact bourgeois determinism does by having all events lying on the table.

How do we know that the shuffling of likes in the future, which is determined, will produce new qualities? Precisely because we do believe in determinism—i.e. that the past was *uniquely determined* by necessary connec-

140

tions. Hence a rearrangement of the past cannot be just the past but must be different (or the connections were not unique). Hence the past necessarily produces a real future (and not a mere oscillation) by its own inner activity. This is evolution.

We saw that a connection between events of a 1-2 character is not uniquely necessary (i.e. one-way) involves discontinuity or 'jumps' and also the production of novelty by a kind of inner activity. Hence 1 and 2 stand in a causal relation. They do not stand so in a Universe containing 'all kinds of numbers' but starting with the premise unity as the only grounds then 2 is the only possible next step. Given 1 and 2, 3 is the only possible unique determined outcome—and so on. . . . Thus this simple relation shows us a causal relation, producing a novel effect in a one-way direction by a kind of inner activity and yet a relation which is strictly determined. It is discontinuous—subject is distinct from object— and yet it is continuous, 2 gathers 1 into itself. The paradox of causality and determinism is resolved. But although the individual items are discontinued, the series of integers as a whole is continuous. It is a series. And it is continuous because it is developed by + 1. At each stage 2 gathers in 1, 3, 2 + 1, 4, 3 + 1, and so on. Hence the series as a whole has as the like predictive basis of its continuity 1. But this is precisely the basis of its discontinuity—the difference between each neighbour in the series. Hence discontinuity is not opposed to continuity, but is an aspect of it.

However continuity is only mediated by 1 for the whole series. For the series 9 — 26 continuity is mediated by 9 and so on but discontinuity by 1. As between the number 10—21 continuity is mediated by 10 and discontinuity by 11.

Hence it is only in the series as a whole that discontinuity equals continuity—in respect of parts of the series they are different.

As the series proceeds it unfolds alternately odd and even numbers—e.g.

$$1$$
$$2$$
$$3$$
$$4$$

Thus 3, which gathers into itself the different integers, 2 and 1, yet inhibits a special quality of 1 not openly revealed in 2, similarly 4, gathering into itself 3 and 1, returns to 2, regaled by 1 to form 3.

Each integer in other words has not only gathered within itself all the qualities of the earlier stages, but also reveals explicitly, as a whole, certain qualities of an earlier stage, but enriched by what has gone before.

In addition however the numbers, as they proceed, form hierarchies. For example in the decimal system after 9 a new domain is reached, the domain of tens, in which 1 now has a different value, while continuing to retain the same necessary connection. This domain it may be asserted is not given necessarily in the unrolling of the integers, but is an arbitrary convenience of symbolization. That this is not the case can however be shown by the following considerations: If there was no such repetition, the series of integers would flow on and would require a different symbol for each integer. Hence the integers would be uniquely determined in a linear order 1, 2, 3, 4, 5, such that it would be completely self-contained. Such a series would exclude the possibility of any other series for any two neighbours would be uniquely interlocking---e.g. 1 and 2. The alphabetical series is of this kind but of course being finite—having a beginning and end—it does not exclude all other series. Such a series would constitute a solid linear Universe, or an object whose inner activity existed only for itself; it would be an object that changed only for itself, that interacted only for itself. It would therefore be completely unknowable; but the completely unknowable species that it does not exist (for even to know that a thing exists is to know something about it).

Hence the formation of domains (10 s, 100 s, 1,000 s, etc.) in the series of integers is not a mere convenience of symbolization, but a necessity for the formation of subject-object relations, or activity, within that series, other than in a purely linear matter. Put in another way, such a series would only have time-like characteristic. For spatial relations it is necessary to form domains. Thus the decimal series not only forms domains of 10's, but also such domains as $\frac{1}{2}, \frac{1}{3}, \frac{1}{4}, \frac{1}{2} \frac{1}{4} \frac{1}{8}$ and so on. These domains, which are multiplicatory or divisory (as the linear series was purely additive or subtractive) form spheres of quality which gather up into themselves the qualities of the former sphere, and make it new. For example, the sphere 10—90 gathers up the qualities of 1—9 in a new domain with new characters. Such a relation is hierarchical or systematic. Put in another way it is abstract, since 10 abstracts all the qualities 1—9. A series of integers without domains would be perfectly concrete, and so could not have relations except of a linear character.

Evidently therefore given a Universe which is spatial, which is not one solid world line, the formation of domains, systems, hierarchies of value and organisms, each of which repeats in a higher abstract and more complex way whole discrete and concrete series below is a necessity of strict determinism. The emergence of systems and organisms does not therefore require the descent of entelechies or forms, but flows from the requirements of determinism in a knowable Universe—i.e. a Universe which does not merely exist only for itself, but in which one part exists for another. Once one admits a Universe in which there is a part (and one does if one talks of self and the rest of the Universe) systems or ordered hierarchies are seen to be necessary. The formation of systems or organisms and hierarchies or 'wholes' which repeat in a more abstract, new and complex way the formation of earlier stages is thus seen to be a necessity of the subject-object relation. Without it there is only the possibility of one solid object without parts persisting in time forming the whole Universe.

But how can an *object* form the whole Universe unless it is determined and defined by what is not-object, by the subject? Evidently the formation of domains involves the persistence of former integers in new arrangements. That is why we said the persistence of like makes prediction possible and why systems are necessary to determinism in a Universe with parts. While we can predict these new arrangements, we cannot predict the new qualities, characteristic of the system, which will emerge. 10 is 1, but it is also the 1—9 system, controlling the whole lower

144

hierarchy and as such it has qualities peculiar to such a system.

Each number is not merely its predecessor $+ 1$, another object plus a novelty, but it can be broken down in various ways. Thus 5 can be broken down into $3 + 2$, $4 + 1$, $2 + 2$ and 1, and so forth. It can only be broken down in terms of its predecessors, but its elements, although the same in quality as those of other objects, will always differ in quantity. For example, 3 and 5 both contain as one possible breaking down, 2 and 1, but 5 has two 2 elements in this reduction, 3 only 1. Hence each integer is historical. This history is not merely the ingathering of previous likes, but the ingathering in a different way to other integers. Each is historically different. Yet this history can be broken up in various ways, each element being reducible to further elements,

(altogetherness of everything)
Dialectic.

until all are bare of quality, but before its breaking-up, forming part of a lower domain—as for example 1,173, broken up into 1,000, 100, 70 and 3. This analysis is abstract because it involves the shedding of subsequent newnesses, whose very emergence is what constitutes concreteness. It is also generalizing, because the lower domains are more elemental and therefore more widely ingathered. Yet the entity as a whole is not merely the sum of the elements because it is a unity; that is, a new thing emerging in time. It is a system, an organism, an entity—all characteristics of domains emerging in reality as a part of being. It is a real development.

The concept of domain involves as we have seen the

concept of parts, of elements, of pieces subsisting simultaneously and under conditions of determinism. Thus the domain 10, for example, comprises:

not added together, not multiplied together, but ingathered into the 10 in a systematic or hierarchical way. But although ingathered, they are not absorbed—10 does not contain the sum of the numbers 1 to 10, and hence the other numbers may either perish or be ingathered into other systems. However, our whole concept of series demands the continuity of existence and thus we see that each of the digits represents the stage in an already continued series of integers. A domain such as 10 is therefore merely a special form of togetherness, of self-contained series, of integers. Of course the domains themselves constitute a series and form domains of domains, etc. Thus a system or domain is merely a special mode of togetherness, of integers, or events in a series.

If any series is completely self-contained, and does not form part with other integers in any domain, it is obvious that it is unknowable and therefore non-existent. Hence every integer must form part of some domain which in turn forms part of another domain. In this way there is a kind of hierarchical or systematic connection of all things. The only domain however of which it must be said that every integer must have connections with it is the Universe itself, which thus constitutes the most general system. Moreover, these connections are not instantaneous, since they cannot be that. The route by which every integer is to be linked with every other may stretch far down the series. The shortest route evidently has a special significance.

It is this universal interweaving of domains, and not the concept of strict determinism as such, which enables us to speak of laws and the universal reign of laws. A law is a domain system. The universal reign of law merely means that every integer forms part of some domain. It does not mean that any one law ingathers all reality. Precisely because a law is most universal, it is the smallest ingredient in the largest number of integers, as for example unity. This does not rob it of its determinative predictivity. The Law of Conservation of Momentum is universal, although it says little more than unity. But precisely because of its universality, it tells us least about quality. Domains make possible abstractness and generalization, and they do so precisely because they delete the greatest amount of newness, individuality and concreteness. Integers concretify in time.

CHAPTER EIGHT

THE mutual connection of likeness in all integers, their highest common factor is the domain of domains; the universal domains into which all things fall, is space. Space is the altogetherness of everything. Thus space is not a matrix of the integers, but the necessary connection between them. It is the likeness and the continuity persistence in integers—the unity. Of course the more universal it becomes the more it becomes bare of quality, a vacuum. It is matter, objects, persistence, cause, substance, abstraction, generalization.

But we saw that the necessary connection between integers in a series was meaningless, was not a connection at all, without an unlikeness, which constituted it a causal relation, as well as deterministic. This unlikeness, this discontinuity, this novelty, this effect, is Time. It is the difference of everything. This Time is not something flowing through the integers, or in which the integers are serially unrolled, but it is a product of their necessary connection. The more particular Time becomes, the barer of quality—an electron. It is activity, experience, change, effect, spirit, concreteness, individuality. It gives necessary connection its uniqueness.

It is obvious that in a series considered singly the passage from 1 to 2 or 2 to 3 is a self-contained passage. As such it is a causal relation with itself—the subject constituting its own object. But this would be unknowable, and in fact we see that each transaction demands the ingression of an outside agent (cause or effect) as for example 2, 1 and another 1 not 4. The 1 and not-1 therefore stand to each other in a causal relation and generate 2. Hence it is impossible for an integer to exist by itself in Time. It can only exist by itself in space. For Time to emerge (an unlikeness) it must have a causal relation with an entity outside it. The altogetherness of everything ensures that this connection 'echoes' throughout all the series of integers. Hence an integer can only exist because it is in a causal relation, with the whole Universe or because it exists for the Universe. The movement of like (unity) about the integers as part of their existence in causal relations is what Time is.

Thus a thing can only exist for itself in space, not in Time. But it is a contradiction to talk about not existing in time and it is impossible for a thing to exist by itself in space, since space is the altogetherness of everything. Hence it is impossible for an integer to exist by itself, not in causal relation with the rest of the Universe as part of a process of development. The phrase is correct however if applied to the Universe of integers as a whole. Since they cannot as a whole be in causal relation with some other thing, the Universe exists neither in Time nor Space and as a whole is unknowable. But this merely means that absolute Truth is unobtainable—the limit can continually be neared. 'God made the integers, man made all else' is correct if we regard God as the integers and concrete matter and all else as abstractions, as particular fields in the integers, as systems or generalizations of a spatio-temporal nature.

Hence nothing can exist for itself. It can only exist as a term in a subject-object relation, of a causal character, which ultimately has connections with all other integers, and generate activity. Knowing is an active causal relation. And as a result of the activity, both subject and object are ingathered in a new quality (e.g. 1 and 1 become 2). Hence there is no unique one-way connection in knowing in the sense that there is in time. Both parties are changed as a result. Knowing is like all causal relations, a *mutually* altering relation of activity. (Every action has an equal and opposite reaction.) Existence is activity.

Since the Universe had no Time or Space, there is no universal time and space. Time and Space do not exist absolutely, they inhere in the relations of the integers. And they can never be separated one from the other even in the simplest relation, since a necessary connection requires an element of novelty to have a meaning. A spatic-relation has no meaning with time, and *vice-versa*.

Hence Time and Space can only mean a particular set of relations of an integer. An integer's space at any (discontinuous) moment of its existence is its universal connections of likeness with other integers, and its time the glow of novelty forming part of those connections. As the integer unrolls itself and the series its connections change, while still remaining most generally like, and with them new novelty emerges. Other integers change round it, and the Universe as a whole develops incessantly. Evidently each integer must have a different time and space because it has a different Universe. For each integer-I the Universe all not-I, and for different integers I must be different, not-I must be different too. Hence Time and Space are relative to an integer and have no meaning for the Universe as a whole, any more than it has meaning to say the Universe exists.

We have therefore established as categories of determinism:

1. Discontinuity of quality and continuity of quantity.
2. The altogetherness of everything.
3. The difference of everything.
4. Existence as the product of the inner activity of a causal relation between integers, of a subject-object character.
5. The universality of domains of quality, which are not however self-contained.
6. The reign of law and its universality—categories of domains.
7. The non-existence of an integer in itself.
8. The emergence of quality because of a subject-object contradiction, such quality ingathering the elements of earlier stages and also exhibiting in an enriched form one particular earlier quality.
9. The relativity of Time and Space as relations of an integer with other integers.
10. The universality and meaning of development.
11. The difference between determinism and causality.
12. The limits and universal power of quantitative prediction.
13. The dialectic of existence (thesis, antithesis, synthesis).
14. The revelation of contradictions which secure development (unity breaks up into unity and not unity 2 into 2 and unity, etc.).

151

15. The meaning of a causal relation and its connection with determinism.
16. The unity of opposites.
17. The meaning of Time's irreversible arrow (flow from past to future).
18. The revelation of all existence as causal activity.
19. The analytical fullness of history in every 'thing.'

All these have been unfolded from the general characteristics of series of integers. But it is evident that they are also the basic categories of dialectics.

At present however they are bare of quality simply because the only difference between one integer and another is mere quality. They are just different sounds. As soon however as we apply them to real matter, to the object, by scientific theory and practice, they become filled with real quality and at the same time of real quantity. They begin to denote reality.

The integers correspond to 'events'—i.e. to real evidence. Each series of integers is an elementary particle. It cannot however exist for itself, but only by causal interaction with other elementary particles. It is therefore forced to form part of domains or systems or wholes which in fact its activity evolves. Apart from these relations, there is no meaning in saying it exists. Hence the existence of every particle is not merely logically but causally determined by every other particle. Merely logically determined particles do not exist—only causally determined particles. Causality is real determinism. Time and space are the most general differences or most particular generalities in these causal relations. The subject is 'Time' (*D* exist, therefore I am). The object is space (substance consists of extension).

It would be a mistake to regard the elementary particles as 'things.' Things in the ordinary sense are parts of a domain, and generally have an elaborate hierarchical structure. A thing does not exist in itself—if it did it would cease to be a thing and become an unseizable individuality. A thing is for example an object—a rock, or stone, or drop of water—in which case it shares a large number of qualities—such as pushiness, shape, length, mass persistence, which are peculiarities of a fairly elaborate domain, that of macroscopic objects. These are already possessed of an elaborate structure and are full of history. The more general one makes one's definition of a thing (almost anything) the wider and less complex the domain. A living thing is a small and highly complex domain. No domain, except the Universe, is all-embracing, and every domain is in causal relation with all that is not-domain as subject to environment. The concept or Idea of Hegel (the form of earlier philosophers) is the domain. The matter consists of the integers or elementary particles. We have seen how the necessity for organization in domains springs from the determinism of the integers. The greater elaboration of domains leads to an increase in activity and the production of quality. Evidently it would be misleading to call an elementary particle a thing and expect it to have the familiar qualities of thinghood since all familiar things are in fact parts of elaborate domains. An elementary particle however is a completely individual entity precisely because it belongs to a most general domain.

How does this view of the Universe, which may be

regarded as dialectical, compare with that of modern physics?

As a brief summary, it may be said that modern physics has been forced by its experimental progress to abandon all the old mechanical interpretations. It has not as yet however found any substitute for the categories which its own research has revolutionized. In abandoning therefore the categories of mechanism, it attempts to use the categories of subjectivism—both bourgeois. Of course the problem cannot be resolved in this way which is in fact a retreat. Or else there is a general feeling that one should be able to do without categories at all—in other words, 'one should do without concepts.' Obviously such a programme is impossible. It is impossible to talk about physics without talking about the electron and the quantum. Thus the field of physics is occupied by opposing armies of bourgeois physicists. Einstein and Planck cling to the categories of mechanism. Jeans and Eddington attempt to find substitutes in bourgeois subjectivism. Dirac and Heisenberg whole-hearted, Schrödinger and de Broglie with less confidence, attempt to do without categories altogether.

CHAPTER NINE

In the world of macroscopic physics there is substantial agreement with the position outlined above, in which the series of integers were used for world-building according to the principle of determinism—the basic principle of science. That Time and Space are not characteristics of the Universe, i.e. that there is no universal Time and Space, but that each particle has Time-like and Space-like relations with the rest of the Universe, is part of the deductions of relativity physics. It also follows that Time cannot be separate ultimately from space, because unlikeness is necessary to make possible a like connection. Relativity physics expressed this in the shape of a space-time continuism. Each series of integers corresponds to a world-line of a particle in the world of relativity physics.

However we saw that everywhere the numbers criss-cross. It is possible there to trace an apparent series in many possible ways. Which is the series-in-itself—i.e the individual particle retaining its identity? It may be distinguished in this way. Between any two numbers there is the greatest ingathering of like. But there will also be a difference which will be the greatest possible between any to integers precisely because the necessary connection between these two is the same. Like is a spatial and unlike a temporal relation. Hence the world-line of a particle, like an indentical integer-series, is that chain of events between which there is the shortest possible space and the greatest possible time—such space and time being the particle's own space and time.

Hence a particle's space and time is uneventful; it proceeds on its way with absolute serenity. Or put in another way, it is at rest. But other particles change their spaces and times in relation to the particle—or put in another way, they move. Because a particle's space and time is at rest, other particles move in that Space and Time. This is merely another way of saying that other particles move. But we saw that to exist in relation to a particle, other particles must have transactions with it—that is, they must have relations, which since they are spatial and temporal relations, must involve distortions of time and space of other particles for a particle. That is to say, to exist for the particle other particles must move in space and time. Therefore in relation to a particle, the existence of other particles is a mode of motion, or *vice versa*. It cannot be said however that existence is a mode of motion for the particle, since it is at rest—or put in another way, it knows only one time and space in which all other events move. Time involves change— the emergence of unlike—hence change is a mode of existence.

It is easy to understand how a particle can imagine there is a universal Time and Space, for all other particles can only be given an existence, a movement, in its time and space.

Remember we are speaking of particles not of things. Things are domains of large quantities of particles, with a complex overlapping and reticulation. They therefore include many times and space, though normally these have great similarities. Thus a thing is tempore-spatially normally a bunch of perspectives of the Universe.

A particle by its own lights does nothing and has everything done to it—or does everything and has nothing done to it. It always follows the only possible unique path (shortest space and longest time) or follows no path at all, being at rest. All this by its own lights is a statement without meaning; but it is evident that these characteristics give it a definite limitation in relation to other particles. Then it can be said to have a path and a velocity. In itself it has neither for its time and space is unchanging. But for any given particles or domain of particles, it follows a straight, curved or accelerated path. It can in fact never follow in relation to another particle a perfectly straight path and even velocity for this would mean that both were absolutely unmoving in relation to the other—in other words that they had the same time and space. But this would mean that these two particles had exactly the same relations to the rest of the Universe —i.e. that 'they' were the same particle.

Every unlike becomes like by being gathered into a new integer (quality becoming quantity). The altogether-ness of everything ensures that this unlike is not all gathered into one world-line but splits off into another for absorption (cause and effect—no entity can be the cause of its own effect). Hence there is what may be called a passage of quality and the shortest possible passage (particulate filiation of qualities) has certain special characteristics. Among these are—a given likeness cannot be ingathered into a more immediate likeness. (No quicker velocity.) It will be a constant velocity in all particle's time-spaces (because in all particle's time-

157

spaces it will constitute the same number of minimum units). Such a relation is the only possible relation between elementary particles. (Its time and space consists of the time and space of the two particles—hence it is a 'blurred' relation later.) It is the straightest possible line between two particles because

(a) to specify a line two points are necessary.
(b) To specify its straightness or curvature other particles are necessary.
(c) These have their own different times and spaces, therefore the path is necessarily curved.
(d) But because transactions involve the greatest possible identification of the two parties spaces and times, it is a minimum curvature.

Now all these various characteristics—lack of absolute space and time, relativity of motion (and relativity of mass and energy, length and shape, which we have not discussed because it flows from it), spatic-temporal characteristics of geodesics and light-rays—are all characteristics of relativity physics. The new relativity laws of space, time, mass, energy, light path, inertia and motion all involve these apparently revolutionary principles, which seem to shatter the Newtonian world-scheme. Why is inertia equivalent to gravitational mass? Because inertia is the 'difference' between the space-times relation-plexi of other particles and *the* particle. Other particles see *the* particle knocked about, changing its direction, etc., and translate this in terms of an inner inertia and an outer force—or an inner mass and an outer attraction. But the particle by its own light is doing nothing; outer things are happening and changing.

We have been dealing with the behaviour of other particles in the spatic-temporal network of *the* particle. In other words we have been dealing with a particle's eye-view of the world—the world as inertia. This is an objective view of all other particles by one particle. But we have not been dealing with the world as seen by one particle; but the world as seen by every particle in the same objective way. Thus it is a world which includes all particles' times and spaces and therefore is unchanging. It is a world without motion—a world of pure geometry—or pure inertia; it is the same. It is impossible to talk about causality, because there is no activity. It is impossible to talk about quality, because the qualities are cancelled out. In such a world events do not happen —one comes across them. There is no causality—only complete determinism. This is Einstein's world, and it is no wonder that Einstein believes in 'strict determinism.'

Einstein of course arrived at his world by very different methods. His is already full of constants—the gravitational constant for example. We have avoided any real constants and merely asked ourselves what would be the characteristics, according to dialectic logic, of a Universe of particles connected by relations of strict determinism.

However there are also differences in our two worlds. Einstein's world and ours do not inhabit the same logical space. They exclude each other. It is our contention that Einstein's world is contradictory because it is still bourgeois and mechanistic—it is not a world of complete relativity.

Our world is discontinuous. It is also continuous, i.e. serial, but there are irreducible discontinuities in

Nature. There are not in Einstein's. In Einstein's world a particle can be located precisely in the space-time continuum—in ours it cannot for locations means locations with reference to the space and times of other particles and they are continually changing. In other words, a particle can only be located accurately in its own space and time, and then this means no more than 'A particle is where it is,' a not very helpful statement. Again, Einstein's world is complete without quality. Any quality that is added is merely supererorgatory—a sort of scum or epiphenomenon. But in our world every event depends on quality and no quality is like any other quality. Hence our whole world is meaningless until the qualities are specified but from the nature of the case, they can only be specified in actual concrete experience. And until experience, our world waits without a *real* determinism. Hence its determinism is not given in itself, but depends on experience. Thus Einstein's Universe contains no real activity, whereas ours is only determined by inner activity of causality.

Moreover Einstein's world is monolithic. But ours is organismal or complex, and evolutionary because of the existence of domains. These are not mathematical domains as such, but domains of quality. A qualitative domain is meaningless in Einstein's world however for this is completely determined without quality. Any domain of quality is thus something stuck on like an architectural ornament.

Now it is a well-known fact that quantum physics negates relativity physics precisely because it postulates a fundamental discontinuity in events and an impossibility of precise location in space and time. It is also the case that experience negates relativity physics because it gives us a direct experience of causality, freedom, power and the emergence of novelty and the reality of quality. It is also the case that biology negates physics because it demands the existence of domains of qualities as a part of determinism.

It is not proposed to deal now with the precise likeness between quantum physics and the dialectic Universe but to pass straight to a consideration of how Einstein came to construct his Universe which was immediately shattered by quantum physics.

The Newtonian and Cartesian world consists of a world of particles in an absolute space and time. This world persists until Einstein.

Now if space and time is something independent of the activity of the particles, it is evident that 'I' who form one of the particles, or a group of the particles, can locate and time the other particles in a way which will be true for all particles. And it is evident that my knowing of the particles will be a knowing in space and time and not a knowing of the spatial and temporal relations of particles; for knowing is a relation between entities, and the relation by 'I' and the other particles by definition takes place in an absolute space and time.

But if that is so, then space and time is unknowable; for we can only know events and events can only take place in space and time. This fluid medium in which all events float, is prior to cognition, because it is the matrix of cognition. It is the unknowable foundation of all that is. Yet this seems contradictory to our experience that objects appear to have substance and change—in other words, space and time seem to be in our apprehension of them

161   M

Now it will be evident that if all the particles can be precisely located in an absolute space and time independent of the particles, then is it possible to know the Universe in a 'Divine' way. That is, one can stand outside it and watch it as one peers into an aquarium tank. Evidently in these circumstances, and given the particular velocities of the particles, one could predict with rigid predeterminism the whole of subsequent events.

Our previous analysis showed that even if this glass-tank theory were correct, one would not be able to embark on a real prediction, because of the emergence of new quality as part of the inner activity of the relations. These qualities are not stuck on already moving Universe but are what the Universe's movement in Time is. Hence complete predeterminism is an illogical concept, since it demands the experience of what is not yet in existence—all the qualities must be lying on the table before arrangement in a unique order.

But in our glass-tank Universe complete predeterminism is possible, because time is by definition absolute, and not generated by the relations of the particles; therefore the Universe of particles develops with strict determinism independent of new qualities, which are only 'stuck on' by experience.

This Universe of Newtonian physics is contradictory in as much as all the particles can be completely located in an absolute space and time—including the particle, or group of particles which is the subject. But a subject's knowing can only take place in a spatic-temporal frame. Hence knowing takes place as it were twice: once as part of the 'I' particle's concrete relations with other particles in an absolute space and time and again when that 'I' particle together with all others and their relations in an absolute space and time are ingathered into a spatio-temporal act of knowing. Obviously this is a kind of infinite regress; and in particular how can the absolute space and time be gathered into the framework of an outside knowing and remain absolute.

There is of course a simple theological solution—the space-time jelly is God. This is in fact Leibniz's solution in which space is the togetherness of everything in God and Spinoza's, in which substance—extension—is ·also God. But if this is the case how can Man gather God—the absolute space-time of the Universe—into a spatio-temporal knowing?

This contradiction of course springs from the special bourgeois relation to the object—Nature—which we have already discussed. Absolute space and time is one of the characteristics of mechanism. Nature is the machine worked by the proletariat; the property which by a simple contemplative knowledge of its necessity, will make Man free. Thus consciousness becomes the mere contemplation of the active object, and theory does not flow into practice as part of one activity. The bourgeois stands outside Nature in theory because Nature in practice has been completely caught up into a class from which the bour-geois is isolated.

Thus, to the bourgeois, knowing *is not a causal relation*. He appears to stand outside the Universe and to know it without disturbing it and, what is more important, without its disturbing him. We already noted how this reflected the whole relation of the bourgeois to society, in which he regards his desires as free, because they emerge 'spontaneously' from the blind market and set to work on Nature through the machine. Thus pure contemplation can affect Nature, the object, and the machine, but these do not in turn affect desire. But this is not a causal relation, for although the cause 'produces' an effect, both parties are changed by the transaction. (Every action has an equal and opposite reaction.) The causal relation is 'unique' only in that the emergence of quality in the form of 'effect' sites the event the right way round in Time. The causal relation with all its activity and power is known in its purity in experience; knowing is full of causality; the complete causal relation requires in the subject knowing as a part of it. Otherwise it is like a shove in the back—we suffer the causal relation; we do not initiate it. In other words, we know Nature only by changing her, and are ourselves changed in the process.

But this means that since change is time, and something must change (i.e. there must be a substratum of likeness, which is space), all spatio-temporal can only be known as a part of causal change. Space and time are sweated out of the activity of particles among themselves. But since no particle has exactly the same set of relation with other particles as any other particle, there is no independent space and time, but only individual sets. And the idea of knowing all particles in a frame of space and time is absurd, since it includes knowing the 'I' particle—in other words the 'I' particle is both inside and outside the frame. This is like trying to lift oneself off the ground by one's bootlaces.

Thus complete predeterminism springs from the special bourgeois view of the object, of Nature as mechanism, and we saw how in turn this springs from the detachment of the bourgeois from the object-changing class brought about by the 'free' market and the restriction of social relations to property rights over objects. Such a predeterminism excludes causality, and thus gives rise to a kind of bourgeois nightmare; in which all one's future acts are predestined. In the more philosophical earlier scientists, this also gave rise to the theological God in whose consciousness were all these movements of the future. Later the scientist—the way being led by Kant—dispensed with this God and put the bourgeois mind in its place. Thus the bourgeois felt his material body, as a part of Nature, insatiably sucked into a predeterministic Universe, where he would—at least in theory—know all his future acts. He became an object to himself, mere property, and therefore unfree and fully determined. This nightmare predeterminism, with all the contradictions it involves, is quite distinct from the determinism we made the basis of the Universe and which includes instead of excluding the causal relation.

It was by no means necessary for Einstein to arise before this absolute space-time received queries from experimental physics. We have already mentioned the conception of ether—how it became necessary to act as a carrier for the light waves and other electro-magnetic phenomena. But evidently this ether is simply the absolute space and time in which all particles float, given a new name.

The characteristic of light and similar phenomena is that it is a transaction between particles. (The immediate neighbourliness of qualities already discussed.) Hence it is the basic spatio-temporal relation—one might talk of it as the atomic spatio-temporal relation if this were not

165

contradictory. Hence that 'in which light travels' is the sum of particles, spaces and times, each slightly different, but in the aggregate making a jelly or wool of relations which has a certain independence of groups of particles, seeing that it is composed of the most generalized basis of each particle's view of its relations with other particles.

Hence 'the ether,' although this was not realized by physicists, was simply a picturesque name for absolute space and time in which all events 'take place.' It was therefore no accident that certain queer behaviour on the part of the ether (the Michelson-Morley experiment) set Einstein on the path which led to the Principle of Relativity.

But evidently there is some distinction. For if 'the ether' is absolute space and time, then light can move in it and have a velocity which will change as the earth's movement through it changes: If however 'the ether' is merely the greatest generalization possible of the various networks of each particle's relations with the others, then light, which is one of these fundamental generalizations, cannot alter its velocity in the most generalized space and time in relation to a particle, because space and time (the ether) is composed of it.

According to this distinction, as experiment showed, the absolutist theory was proved incorrect and the relativist theory proved correct. And thanks to the existence of the tensor calculus, Einstein was able to show by a very beautiful and ingenious development how the ether is composed. I do not propose to go into detail into Einstein's argument here, but briefly what his use of the tensor calculus boils down to is this. Every particle's possible relations with all other particles are taken in a mass, and 'thinned out.' All those relations which are peculiar to the particle are 'thrown away' and we are left with a number of common relations, which may be called world-invariants, and are the most generalized relations possible to particles. Since each particle groups its likes and unlikes in a different way (i.e. has its own space and time) such a generalization cannot consist of something separable into space and time, but must contain them firmly interwoven. With any one particle's world-view one can draw a line or rather cross-cut and say—before this all qualities are known and like—after this they emerge. This is not however possible with an amalgam of space-time, and hence this generalized invariancy is a fourth dimensional continuum. This is Einstein's Universe.

But evidently if we regard it as a real world-picture we are only back at the old bourgeois error—although admittedly an error at a higher level. For although space and time are in this view only relative, Einstein's space-time continuum and its generalized intervals are taken as absolute. All events happen within them—all particles move inside it. And now we see the basic error is similar to that of the Cartesian. The knowing particle is included 'inside' the Universe, and yet is supposed to be able to mediate universal relations. Hence this Universe is even more erroneous, or rather abstract, in its formulation than the earlier one. For every particle's world-view is included in the generalization, and each particle's world-view must include the knowing particles as known or (once) as knower. The method hopes as it were to cancel out the knower, but in fact it only adds him up, and gives a Universe of pure knowledge, pure contemplation, pure geometry. The continuum is merely the old ether.

167

Thus the Einstein Universe is an abstraction from concrete living—the greatest possible abstraction. Hence its power, and also its limitations. The most definite limitation is its lack of causality and hence its lack of inter-atomic relation. Its particles are all identical and therefore quiescent.

Such a world expresses bourgeois determinism and mechanism in its highest form.

CHAPTER TEN

*Zeno—The calculus—Quantum (opp. light. All par-
ticles view of one)—Frequency and discontinuity—
Continuity and discontinuity (unity)—Entropy—
Accident and Necessity—Freewill—General to
science*

EVIDENTLY Einstein cannot believe in freedom or the causal relation if he takes his causal world as absolute, as he evidently does.

There seems no doubt that Einstein's world represents the final productive development of the bourgeois world-view—Nature as the object in pure contemplation. It is the climax of mechanism. For almost at once it came into irreconcilable collision with atomic phenomena, the more irreconcilable because of its greater penetration and generalisation.

Yet in fact the problem with which quantum physics faced relativity physics was one of the oldest known to physical thought, although it was a problem enriched by all the subsequent experience of physics. It is the paradox of Zeno regarding motion. For example, Achilles and the tortoise. Achilles moves twice as fast as the tortoise and starts separated from it by a distance X. When Achilles has covered X the tortoise has covered $X + \frac{1}{2}X$, when Achilles has covered $X + \frac{1}{2}X$, the tortoise has covered $X + \frac{1}{2}X + \frac{1}{4}X$ and so on, so that the tortoise always covers a slightly larger distance than Achilles at any given time. Achilles therefore can never catch the tortoise; but we know that in actual life Achilles can catch the tortoise, and that when Achilles had covered $2X$ the tortoise has only reached $\frac{3}{2}X$ from Achilles' starting point and Achilles must therefore have passed him.

Now these and the associated paradoxes were used by Zeno to prove that motion was an illusion, because it contained an essential contradiction. The moving object, the flying arrow, both were and were not in one place at the same instant. Subsequently philosophy never satisfactorily tackled this problem. For example to say as Bergson does that the line of the arrow or of Achilles movement is only an abstract trajectory and not concrete motion, which cannot be presented in this way, does not explain what concrete motion is. The problem formally is solved by the differential calculus; but the calculus does not apply to atomic phenomena. In other words continuity is a characteristic of macroscopic phenomena, but if we analyse sufficiently we reach point-instants (as in the tortoise and arrow problems) which are discontinuous. Discontinuity is therefore not so much a discontinuity in space and time, which would be unthinkable because space and time are generalizations, that is broad continuities, but a discontinuity of the particles of matter, to which space and time adhere.

Space and time are macroscopic relations. They are relations of continuity simply because they are the most generalized aspects of likeness and unlikeness in the world-relations of particles. And the most generalized aspect of space-time is motion. 'Distance' is a purely like and spatial relation. But motion involves change and therefore the ingression on unlikeness—time—although it is the most abstract and general ingression possible, change in spatial relations—change in general likeness. Pure distance is in fact a meaningless conception, because it is unknowable. Distance can only be known by the motion of something between relata, for physical distance involves a physical relation. Such a relation must be motion. In this sense distance is secondary to motion; motion is a prerequisite for spatial relations as such. Motion is therefore existence; the contradiction rooted in it; the inner activity of what constitutes the space and time in it, is the existence of the particle.

But motion itself can only appear in its most generalized physical form; and this proves to be the relation of the neighbourliness of particles—the passage of a light ray, *interval.* Zero interval is the main generalization of the physical Universe and it is therefore the stuff of continuity. Precisely because it is a generalization it is continuous— but its true continuity depends on its universal generalization. It is therefore only *absolutely* true of the Universe as a whole. Continuity is simply an aspect of the Universe considered macroscopically, as a whole, as a unity. But a unity is in fact the ultimate discontinuity. Continuity is not therefore exclusive of unity, although it is opposed to it. Absolute continuity is identical with absolute discontinuity.

This shows us why the Einstein Universe, in spite of its purity and generalization, breaks down directly it comes to a real relation of knowing. It is only true of the Universe as a whole—as a super generalization including every particle. But any relation of knowing implies one particle as detaching itself from the rest of the Universe as object or subject of knowledge. It implies a causal

171

relation between a particle and others, which because it involves the emergence of an effect, is a unique relation, with a one-way arrow, which at once involves a discontinuity in the whole system which must from its very essence escape the network of super-generalization. Thus the Einstein generalization is at once contradicted by the most elementary relation of knowing possible in the Universe—the ascertainment of the position or velocity of the Universe. It proves to be hazy by a certain definite blur or overlap, which blur or overlap exactly equals in dimensions the minimum quality possible to a causal relation—the neighbourliness of integers—a quantum of action.

Hence it is not true to say that space and time and the passage of light rays and the motion of particles in space-time are discontinuous, for space and time and the path of a light ray are generalizations which are only true of the Universe as a *whole*. But any physical relation of a causal character involves the Universe *not a whole*, but knowing itself in a relation of a causal character. It would be equally fallacious to suppose that discontinuity is primary and continuity and determinism a mere approximation to a limit. Both are abstractions, one a generalization and the other a particularization from real being. After making the first mistake, physics is inclined to swing over to the second.

Einstein's Universe knows nothing of action, because action involves activity, the causal emergence of change, and this is a matter of the subject-object relation. Action does not involve that the bourgeois steps out of his Universe and yet continues to know it in a frozen contemplation; frozen because all causality, all discontinuity, all multiplicity has been removed from it. Action is however the main category of causality physics, because the emergence of each quality involves a quantum of action. Einstein's world is pure geometry but the world of atomic physics is a world of integers. Relativity physics studies absoluteness—necessity, determinism—Quantum physics studies freedom, causality, relations, activity.

But if relativity physics, having constructed its world of absolute continuity, immediately finds it shattered by the most elementary relation of knowing, so does atomic physics, having constructed its world of absolute discontinuity, find it shattered by the same elementary relation. The Universe of physics springs from abstract knowing, abstract knowing springs from concrete living —Man's real struggle with Nature, and the more completely any ideological Universe attempts to detach itself from the foundations, the more shattering is the explosion as the irridescent bubble is torn away from its source.

The contradiction against which quantum physics has shattered itself logically (while remaining intact as a practice, as an experimental technique) is the question of whether an elementary particle—of which for the moment we will take the electron as typical—is a particle at all. To take a well-known example. A quantum of action is emitted by an atom in the form of 'light waves' and these spread outwards throughout the Universe. They then strike an atom which can only receive a quantum of action—all the light waves, dispersed all over the Universe, therefore immediately 'vanish' into the absorbing atom.

173

*Light waves as quanta—Atoms as waves—Integers
of frequency—Heisenberg's matrix—Schrödinger's
wave equation—(Probability waves 4. 3 dimensions.
All waves: probability)—Atom as object (quanta)—
Atom as subject (light waves)—Subject-object relation of knowing—Entropy—Accident and necessity
—Freewill—General outlook—Mechanism (not just:
but bourgeois)—Subj./obj.—Practice*

Evidently this is an impossible kind of conception
One way out of the contradiction is to accept it just as it
stands; to regard the emission of light as subject to wave
laws and the absorption of light waves as subject to
atomic laws. But evidently this is no way out of the
contradiction at all, since it involves its acceptance just
as it stands, without proceeding to a resolution.

Our previous analysis ought however to give us a clue
to the nature of the contradiction. We saw that light
'rays' were the neighbourliness of qualities—that they
were the most generalized and abstract relations possible.
Light waves in the ether—the rules governing the emission
of light, are therefore waves in the ether, which, as we
say, is simply a name for absolute space and time—or
rather, since this is the most generalized space and time
possible—a name for the spatio-temporal invariancy of
the Universe as a whole. This in turn is only another
expression for all in the Universe which is not-matter—
i.e. not particulate matter.

Such relations however can only be true as relations of
complete continuity, true of the Universe as a whole.
Hence as long as a light ray is pictured as remaining in
the ether, it is pictured as participating in all-universal
relations of wholeness and continuity. In the Einstein

174

theory, or again in the earlier ether theory (both of which postulate an absolute space time continuum) the waves of a light quantum are pictured as spreading to every part of the Universe from their centre.

But as soon as the light quantum is absorbed by the atom, it becomes part of a causal relation between the two atoms. As such it involves a fundamental discontinuity in the whole Universe. The Universe is split in a subject-object relation of a causal character between emitting and absorbing atoms. Such a discontinuity must at once shatter the framework built up on the basis of a complete continuity.

In fact it amounts to this; are we interested in a part of the Universe or the whole? If we are interested in the part, then this involves causality and discontinuity; if in the whole, determinism and continuity. But evidently either interest involves us in a contradiction. If we are interested in the first, we must be as knowing particles either subject or object of the causal relation—we can never know it in its fullness and must introject our split experience into Nature. If we are interested in wholeness, we can only know the whole that is not us, and therefore are bound to introduce a veiled discontinuity; we can never know the Universe including us, any more than we can lift ourselves off the ground by our bootlaces.

We can never know a light ray; we can only know a quantum, either absorbed directly, as when we see a star, or reflected from a mirror, or from suspended droplets in the atmosphere through which a ray passes. Light is a causal relation between atoms and therefore the tracing out of light waves in the ether is the tracing out of a relation independent of relata. As such it has no physical meaning; it can only have a symbolical meaning. We know that in a subject-object relation of knowing it is possible to take the subject as real—idealism—the object as real—mechanical materialism—or the relation alone as real—positivism. In the latter case phenomena become a kind of screen between object and subject. This is exactly what light has become as a result of the development of the wave theory. No doubt the invisibility of air and its capacity to reflect and refract light has had some responsibility for this tendency to regard particulate atoms as floating in a sea of light, like rocks in a breaking sea. This conception of course gives knowing and perception a positivist air which is reflected in idealist philosophy. Perception is not an active relation between eye and object (except in Cartesian philosophy) but a kind of fumbling of fluid at the gates of sense. This fluid seems to carry clues to outer reality, rather than immediate knowledge. We are lonely islets bathed in the etheric ocean:

> 'Yes; in the sea of life existed
>> With echoing straits between us thrown,
> Dotting the shoreless watery wild
>> We mortal millions live *alone*.
> The islands feel the enclasping flow
> And then their endless bounds they know. . . .'

But in fact the outward spreading ripples of light are fictive, as the quantum theory shows. Light cannot spread out in space and time, for the simple reason that it is the stuff of space-time. The neighbourliness of qualities, from which the universal generalization for the universal continuum is extracted, is a light ray. Hence space-time is extracted as a generalization from light rays. While therefore the relativity categories are true for the Universe as a whole, they cannot possibly form a matrix for the particularities of which they are the generalization, any more than specific qualities can contain inter-specific difference. John, James and so on belong to the genus Man—Man is the generalization of their characters. But John's club-foot and James's dark hair are not specifically human characteristics. This contradiction between species and particulars, between nominalism and realism, between fôrm and matter is elementary to a description of a whole Universe from which they are merely abstractions.

Hence the whole structure of light waves in ether must be dismissed as purely fictive, and symbolical. A light wave can never be known—only a quantum, or whole atomic quality. How then can these unknowable light waves be said to exist? The contradiction between the rules governing the emission and absorption of light can only be solved by deleting the whole apparatus of ether with its waves.

Certainly no concept of science could have a more furtive history. It began as a fluid to support the planets; it next became a rigid bearer of waves; and then revealed itself to have no motion relative to the earth. Its qualities then have always been non-material and *sui generis*; which is to say it has been the carrier of all the misunderstandings and incorrect formulations of physics. In sweeping it away, we sweep away one of the oldest problems of physics—what is it that waves, since it is not matter and has rigidity?

However the phenomenon of diffraction reveals a

177   N

visible wave pattern; and even if wave patterns were not clearly visible in light effects what meaning are we to attach to the complex mathematical apparatus based on the experience of waves in matter?

It is becoming increasingly plain that they are not real waves but mathematical waves. This does not mean light is a wave-equation. On the contrary, it means that what we thought as a result of the lectures of the physicists was really light—the waves in an immaterial ether —is nothing but a mathematical abstraction and that light is *really*, exactly the causal relation we experience when we see a star. So far then from the abandonment of the ether to a mathematical domain involving the dematerialization or mentalization of phenomena, as Jeans suggests, it in fact means its concretization.

The waves of ether are probability waves. If we regard the atom as a point centre of activity towards the rest of the Universe as a whole, then we regard the rest of the Universe as 'generalized.' Hence there is an equal probability of the object of activity being any part of the Universe. In that case the light ray could be represented as a fog spreading equally all over the Universe at the speed of light.

But we know that the atom is more than a point centre of activity; it is a particle, a hole in the continuity of the Universe. This imposes a certain limit, the quantum to its emission of light and this quantum, cutting across space-time, can emerge as light of various 'frequencies' or wave-lengths. Hence the fog becomes a wavy fog—the orthodox etheric light wave of earlier physics. But already in this conception of the quantum, the causal relation has in fact been furtively introduced, for the quantum can only exist as a quality emerging in a complete subject-object relation. Hence the object of the causal relation already exists in the wave-fog in a veiled form. It is the shape or frequency of the waves.

Directly the complete causal relation is openly envisaged, which is expressed symbolically as the fog-wave reaching the absorbing atom, then of course the whole probability wave system disappears. The light waves all over the Universe vanish. The probability has become certainty. Only now does the quantum emerge as a concrete quality—or rather, only now has it emerged, for since its emergence takes place in space and time, it must occupy space and time. Thus we see that both the ether wave theory and the quantum theory are partial aspects of a complete causal relation—a subject-object relation. The ether picture is an idealistic picture: it is a picture from the point of view of the subject—the emitting atom, in which the subject alone is taken to be real. It is an idealistic conclusion and taken to its logical conclusion makes the light relation a mere equally dispersed fog of probability.

The quantum picture is a mechanical materialistic picture: it is a picture from the point of view of the object, in which the object alone is taken as real. The quantum is firmly attached to the atom considered as an object, instead of emerging as a quality accessible to knowledge from the causal relation. The contradiction can only be resolved by a conception of the subject-object relation as a complete active relation.

A probability wave seems at first a ghostly concept. But in fact all waves are waves of this nature. In a wave

179

on a pond the water does not actually travel, although the waves move from one side to another. What is it that travels? It is futile to answer, the waves. It would be truer to answer: the motion of a motion. The particles of water reciprocate, but there is a something in the motion that travels. This something is a probability of the particles being at certain heights.

At level *A*, the probability becomes zero, at level *B* it becomes very high. The 'gradient' of probability therefore describes the shape of the waves, and this gradient moves. Hence we see that probability has a special meaning here and that it does not mean that the motion of the particles in themselves is undetermined, but they have a flux or fluidity in relation to the field of the spectators which makes them a wave system. The individual molecules do not trace out waves, and their motion can therefore be regarded as determined by a differential equation and not by a law of probability. But the *molecules as a whole in relation to the space-time field of the observer* do form waves. Probability waves of this character are therefore spectorial domains. They are new qualities emerging because the molecules of water form a specific relation within the water-observer domain. They do not belong to the molecules or the eye, but to the whole relation between the two.

But it may be said that light waves are different from water waves. Water waves are waves of water—of matter. Something is there to act as a 'carrier' of the probability. In the probability waves that are light rays, nothing exists to 'carry' the probability, since we delete

the ether, and thus these waves do become concepts of a
novel ghostliness.

This however is not correct: it is a remnant of the old
mechanism of physics. If we first erect the ether and then
delete it, we certainly get spectral concepts. But if we
avoid this mechanistic conception, then we see that light
waves are waves of matter, just as water waves are waves
of water. They are probability waves for this reason that
just as the molecules of water do not wave, so the atoms
do not wave—but they move in such a fashion that their
motion can appear in the observational domain as a
light wave, just as the movement of the water particles
can appear in the observational field that waves. Light
waves are the motions of a motion. Hence the ether in
which they 'wave' is simply a name for the space-time
field of the observer (or object particle). It is the rest of
the Universe as seen by the observer. In such an ether
the waves necessarily are probability waves: they represent the probability of the distribution of light quanta,
not because there is no determinism about the light
quanta themselves, but because just as a single molecule
cannot in fact wave, so a single quantum cannot wave.
The wave concept has a physical meaning as a macroscopic concept, as an aspect of the behaviour of a number
of quanta, just as the water wave concept has a physical
meaning as an aspect of the behaviour of a number of
quanta. Hence probability does not apply to the
behaviour of a single atom: only determinism can apply.
Probability is a quality of the determined movements
of numbers of particles as they enter a new domain—for
example an observational field. The togetherness of the
water molecules in a specific observational field is the

181

domain of probability; it is a new quality emerging; the motion of a motion. Hence a wave system is an elementary example of the emergence of a new domain, of a new quality.

No amount of calculi about the movement of a single water molecule can give us certainty about the shape of the wave for the reason that single molecules cannot wave—they can merely oscillate. Only molecules together in a special way can wave, and then the determined motions of their single particles become subject in the new field to probability laws. But when the Universe in its wholeness is taken, there is no longer a possibility of being together in a field, for all particles are included by definition. Hence once again probability gives place to certainty, as in the Einstein laws.

We thus see that the case of the single particle and the case of the whole Universe are cases to which the laws of strict determinism exclusive of accident apply. They are also cases in which continuity and discontinuity have the same meaning. They are also cases which exclude the causal relation. In other words they are limits of absolute certainty.

But both are unknowable cases. To know the single particle it is necessary to have a causal relation with it, in which case it ceases to be the single particle. To know the Universe it is necessary for the particle which is oneself to detach oneself from the rest and enter into a causal relation with it, when it, the object, no longer remains the Universe as a whole. Hence the two cases of absolute certainty, the single atom and the complete Universe, which represent as it were the two goals of physical certainty, are unknowable non-existent goals.

This does not however mean that science cannot come nearer and nearer to them by laws of increasing probability.

Thus accident (or probability) is not exclusive of necessity, it is an aspect of it. Accident is a domain of necessity—the causal domain. The necessary determined movement of the water molecules becomes a probability wave when they enter as a whole into the observational domain of the spectator. Accident, or probability, is the emergent quality of domains.

But this in fact was exactly what we would expect from our initial study. For we saw that to have a necessary connection between 1 and 2, a connection of the ingathering of like, it was necessary also to have an element of unlike, for it to constitute a connection. Hence for any domain to constitute a domain, it must have a sphere of qualitative accident, which is the aspect of its necessity in this domain, and constitutes the domain. Thus the peculiar qualities of life are all accidental. They are however all tied securely to the biophysical basis, and in addition generate laws among themselves which are the specific laws of biology.

It is therefore no accident that light quanta make diffraction patterns, just as molecules make wave patterns. They make it so to speak without knowing it: it emerges as a pattern only in the eyes of an observer, just as the blind limited strivings of individual capitalists in society reveals itself as the ordered evolution and explosion of a whole system of productive forces. It is this sense in which history is accidental. A retina or a photographic negative consists of a broad area of atoms and the probability of the quanta spreading

183

themselves in a defined way over the area are what the waves are.

When we say that light consists of waves in matter we mean it literally. In this sense light, like all other phenomena, is a mode of motion of matter. But it is not the motion of molecules in an oscillatory manner like heat: it is the motion of motion systems, like wave motion in matter. Of course the wave-lengths of light do not imply a physical to-and-fro motion of particles of that scope any more than water molecules oscillate to an extent indicated by the distance between troughs, or the material molecules in compression waves of sound to an extent indicated by their wave-length. Nor is it surprising that light velocity is a constant—because it is a characteristic of the most generalized observational field. The observer carries with him the field in which the waving takes place. This is not so with water waves, because we 'see' waves: in other words the waving does not take place in our observational field directly, but it is a waving of waves of light—macroscopic waves built up of microscopic waves. The change which takes place with increase of motion by the observer is somewhat different, and produces a general alteration of the wave shapes.

Hence the main difference between water waves and the particle waves which are light, is that the water waves, as a macroscopic phenomenon, take place in a space-time field, whereas the light waves cannot do so, because they are elementary, and constitute the space-time field.

It will be realized that this conception of the nature of quantum relations is built on Schrödinger's wave-mechanics. Schrödinger however was primarily concerned with another problem; that of explaining the wave character of electrons. Hitherto electrons had been thought of as discrete particles and in certain respects they continued to behave as discrete particles. In other cases however they exhibited diffraction phenomena, which is 'wave' behaviour.

This is the reverse of our preceding problem: How could a light wave spread throughout the Universe, and then suddenly disappear into the atom? The problem now is: how can an electron be everywhere in the Universe and then suddenly appear in one point as a discrete entity? But once it is posed in this way, we see how the problem arrived. Huyghens and Fermat constructed a wave theory of light, which proves to be contradicted by a particulate behaviour. Rutherford, Bohr, and Planck construct a particle theory of matter, which proves to be contradicted by a wave behaviour.

Evidently however the problem has only arisen because of the initial separation of space and time from the things that happened in it. Once space and time were dragged away, they dragged after them light—the most elementary forms of happening, and must inevitably drag after them the things to which happening occur, so that ultimately the whole Universe becomes waves in space-time. But they cannot be waves in a real space-time, for relativity physics has already unearthed the contradictions that lie in this one space-time. There can only be the aggregate of space-times, for all particles. But Schrödinger's is a theory, not about the Universe but about particles: hence each particle must have its own space and time. Schrödinger therefore prefers to call this a configuration space of four dimensions, in which the electron is represented by probability waves.

185

This configuration space is filled with an evenly diffused mist of probability: but as other electrons enter it localization occurs: finally the mist is concentrated at one point—this is certainty. That therefore is the electron. Throughout however the waves have only been probability-waves.

But actually in our picture of a light wave, we could have done the same. We passed straight from the wave spreading out to the Universe to the wave finding its object and disappearing (instead of appearing). In fact however if in our picture we had gradually put in the particles in their real positions and states, the symmetrical light wave system would have had to be altered by minor turbulencies and cancellings out of the waves, because an atom has to be at a certain state of frequency and excitation to absorb a light wave and the 'effect' of this on our map would have been to make large areas of the spreading wave impossible—i.e. waveless in that particular area, till with the last stage the wave becomes a quantum and vanishes in the electron. Obviously with a light wave that reaches to the confines of the Universe before absorption, every particle must be 'put in' before the final certainty is reached. In any case the positions of the initial particles depend on those of the rest; hence for exact specification all the Universe is required in any case before certainty is possible. Hence it becomes obvious that the light wave is only a way of expressing the togetherness of the vibrating atoms of the Universe in relation to the oneness of the emitting particle. As such it has the same physical reality as a water wave; but it has also the same material unreality and mere probability, for it is a togetherness in relation to one particle. Hence although theoretical certainty is attained when all particles are known, since a light ray can only be known as the absorbing particle, as the final object of the causal relation, not all the particles can be specified, because one of them is the knowing particle. This difficulty has cropped up several times.

Thus the light problem, as we propose its solution, is an exact inverse of the particle problem, as Schrödinger proposes its solution. This exact inversion cannot be a chance—the quantum enters both and in fact it is fairly evident, once it is attacked from this angle, that they are the same problem.

For we can only know electrons causally—that is, linked together by light or electric-magnetic waves of some kind. Hence we always stand ultimately to an electron in the relation of particles linked by a 'light ray.' But if this is so, light apart from particles, or particles apart from light rays, are unknowable. There is only the complete subject-object relation: light when it disappears into the atom becomes matter: matter when it appears as an electron becomes light.

Hence the quantum-wave antinomy, with its apparent antithesis of discontinuity and continuity, causality and determinism, accident and necessity, is the result of separating the subject-object relation in its concreteness. A particle regarded as an object in itself, as the object of perception exhibits discontinuity phenomena. It is a discrete particle. The quantum is the stuff of this particularly. But this relation is not complete, for the object is known by something in the rest of the Universe and hence this relation is ultimately determined in its exact shape by every other particle in the Universe—hence it

187

becomes a relation set against the background of the Universe as a whole—i.e. a generalized relation of continuity and therefore the particle becomes a localization of probability waves.

In the same way if we consider the emitting particle in itself, as a subject, as a cause (ourselves seeing its emitted light) it quantizes. If however we consider the relation against a universal background (as we must to complete it) it becomes the centre of a spreading ripple of light.

Hence out of the complete subject-object relation in itself we can extract—the elementary subject in itself (the photon), the elementary subject as located in the Universe (the wave). The elementary object in itself (the particle), the elementary object located in the Universe (the wavicle). Of course if we consider the relation as one of knowing, we become, though subjects, the object of the causal relation and the descriptions must be reversed.

This essential difference between the subject-object and object-subject relations is I think expressed in Dirac's somewhat mystic equation.

Thus Heisenberg's Principle of Uncertainty and Schrödinger's wave-mechanics are aspects due to a cleavage of the prior subject-object relation in its active entirety. Hence the emergence of contradictions. For we have seen that the particle-in-itself and the world-as-a-whole are both unreal abstractions. One is the foundation of Heisenberg's physics and the other of Schrödinger's. Hence with Heisenberg one seems to get accident without necessity and in Schrödinger necessity without accident. Heisenberg's particles are causes but do what they like; Schrödinger's are strictly determined but do not cause and therefore do not exist. We have seen however that accident and necessity, determinism and causality, discontinuity and continuity are not in fact mutually exclusive but a unity of opposites.

Either—or: Entropy—Freewill and causality (Einstein: Planck: Eddington, Jeans)

In his Nobel address of 1933 Schrödinger compares his wave theory with the earlier particle theory as follows:

Either this or that (Particle mechanics)
(aut—aut)

and

This as well as that (Wave mechanics)
(et—et)

but in fact this is not really the situation. By reducing all matter to the movements of particles in themselves, the earlier mechanics successively generated the contradictions of the ether, the wave theory of light, the continuity of relativity shattered by the discontinuity of quanta, the photon behaviour of light and the wave behaviour of particles.

But equally by reducing all particulate behaviour to waves, Schrödinger chooses the other side of the antithesis. As Schrödinger himself admits:

'From the standpoint of wave mechanics the innumerable multitude of possible particle paths would be only fictitious and no single one would have the prerogative of being that actually travelled in the individual case.

189

But . . . we have in some cases actually observed such individual tracks of a particle. The wave theory cannot meet this case, except in a very unsatisfactory way.

'The contradiction is not thus resolved. It is simply taken from another angle—or, instead of either. The difficulty depends on this—"the light ray, or track of a particle, corresponds to a longitudinal continuity of the propagating process (that is to say, in the direction of the spreading); the wave front, on the other hand, to a transversal one, that is to say, perpendicular to the direction of spreading. Both continuities are undoubtedly real." '

But this contradiction is resolved when we grasp that the longitudinal particle track describes a relation between quanta—it is thus subject to exact prediction, although this prediction must necessary be discrete, because a complete quantum is involved. This certainty is governed by Heisenberg's principle of uncertainty. The uncertainty is only such if we regard the relation as taking place in space and time; but since in fact space and time is woven of quantum relations, such an uncertainty is meaningless. Hence particle tracks, governed by Heisenberg's Principle of Uncertainty, represent atomic determinism, because the relation is completed.

The transversal wave front however represents the relation while it is taking place in space and time. Since quanta overlap, the idea of such a causal relation taking place in space and time has a meaning—it takes place in other particles' spaces and times—in the observational field of the rest of the Universe. The travelling wave front, rippling throughout the Universe, then represents the probability of the relation being consummated at any portion of the field. As soon as the relation is completed, the probability becomes certainty, the wave front vanishes, and a quantum appears.

Hence until the relation is completed, probability is the correct description; afterwards this gives place to certainty. Hence Schrödinger's deterministic and continuous Universe is a Universe of probability, and Heisenberg's discontinuous and causal Universe is one of certainty. This is exactly the opposite of what is generally supposed. But it is evident that since Heisenberg refuses to deal with anything but observable phenomena (matrix mechanic) nothing enters his world which has not already happened. Now anything that has happened is certain; and since every happening is a discrete quality, such a Universe is necessarily discontinuous. Such a Universe is not however uncertain therefore; it is a recorded Universe.

Schrödinger's Universe however deals with happenings that are uncompleted. It is a predicted Universe; a forecast. Hence it is only probable; for nothing is completely certain until it has happened. The probability itself is however certain—it is given that logical form—and hence Schrödinger's Universe appears to be deterministic—but it is a determinism of probability. Moreover it is generalized and universal because there is always a possibility, until it is consummated, that an event may turn up anywhere in the Universe. Odds of 6 to 1 are a probability —but the odds themselves are not probable—they are as certain as any deterministic law.

Now evidently science must observe and it must also predict—it must predict to observe and observe to predict. Hence Schrödinger's and Heisenberg's Universes flow

191

into each other. In my opinion, Schrödinger's Universe should be regarded as fictitious; the fact that interference wave bands are actively observed is no criterion. If the probability of the presence of a quantum from a certain source increases from the top of this page to the bottom, then a shower of quanta will produce a characteristic shading. And since the discrete character of elementary relations gives the probability a wave distribution, then the characteristic shading of wave bands must be expected. The behaviour of water molecules is only a probability until the waves have passed, it then becomes a certainty.

Hence we see the importance of the observational domain in producing probability wave systems. Probability requires a space-time perspective. Until an event is consummated, while it remains in the future, it can only be probable, the exact probability being the exact amount of like in it—the quantitative predictive basis, which is itself quite clear and defined and necessary. As it emerges, the not-like is revealed in it as accident, as aspect of necessity, as an effect of a necessary cause. But since the event is consummate, it becomes subject to determinism; the accident is ingathered into necessity, the unlike quality becomes quantity; and the whole relation is one of complete determinism.

But all this has taken place in an observational field; with water waves the spectator's field. We saw that considering the relation in itself, as a pure causal connection between atoms, there is no possibility of wave movement in a domain, for a domain is excluded. The quality is itself the stuff of space and time and cannot therefore move in space and time or have a past or future. Hence it emerges as a discrete entity. The event remains purely causal.

CHAPTER ELEVEN

THIS should throw some light on what is emerging as a principle of first importance; the law of Entropy. This is now regarded by physicists as the primary law governing all physical happenings. It has however a statistical basis and certain novel features which distinguish it from older 'laws' of Nature. However it becomes less odd when we bear in mind our own conception of a law as a feature of a natural domain—or more correctly, the specification of a natural domain. The Law of Entropy is a probability law and yet it is regarded as the basis of the determinism of physical processes. This again is not surprising, since we have seen that accident is an aspect of necessity.

The penetration of science is not based on linear determinism and causality (i.e. that quantum to quantum necessarily succeeds) but to hierarchical determinism or the necessary connection between domains and domains of domains. In the same way the interesting accidents or qualities of Nature are attached to domains. Hence the importance of natural 'laws'—physical, biological, and chemical.

The law of Entropy is claimed to be the only physical law which gives Time an arrow—which has an irreversible character. By virtue of the statistical method the law of Entropy has taken on the following content; every process or event proceeds from a relatively improbable—that is to say, more or less molecularly ordered—state to a more probable one—that is to say, to a state of increasing disorder among the molecules.

I regard the correct understanding of the law of Entropy fundamental to an understanding, not merely of physics, but to the whole development of reality. It is a physical evolutionary law and as such is the foundation of all higher evolutionary processes.

If we take matter completely evenly dispersed in a kind of nebula, we have complete order. Every particle is precisely in a unique place. Nothing can be more orderly than a blank whitewashed wall without a stain or a smut. But such a Universe is a Universe of complete disorder, for nothing can be imagined more completely shuffled than a nebulosity of particles. Consequently in the sense in which Entropy uses order and disorder, such a Universe is completely orderly and completely disorderly. Any movement would disturb the order; any movement would lessen the disorder.

Such a Universe cannot be regarded as a beginning, because beginning is in space and time and in such a Universe all the relations which generate space and time have ceased to have a meaning. Space and time does not exist. Evidently such a Universe is a kind of abstraction.

So far from being stable, as might be thought, such a Universe is completely unstable. For any movement of any particle at once disturbs the state of 'utter nebulosity.' Continuity and discontinuity have precisely the same meaning in such a Universe but directly a movement has taken place, something specific has entered the Universe. The relation introduced by the movement, whatever it is, forms a kind of axis round which the whole Universe now becomes orientated. For example, if it is

194

a causal quantum relation there is a separation of subject from object, and a further separation of causal field from universal background. In so far as the monolithic unity of the Universe has given place to a domain, to a complexity hitherto non-existent, this may be regarded as an increase of order—as the disorderliness of the completely shuffled Universe being subdued by an initial hierarchy which, in spite of its simplicity, is of greater complexity than the original monolith.

At the same time the cracking of the monolith involves a certain disorderliness, a lack of homogeneity, of the Universe, which means that universal order has been broken down by a particulate outcrop.

This first relation which as it were cracks the monolithic nebulosity in two may be regarded as a quality, as the emergence of the new, as accident. It is the product of order passing into disorder and also of disorder passing into order.

Those familiar with the Hegelian dialectic will realize the parallel in this physical picture to the logical first stage of the dialectic. Hegel bids us think of being in its most generalized form, emptied of all particular qualities, as being in its essence. But being in its essence, bare of quality, is indistinguishable from its particularity, its determination, not-being. In the same way essential order is equivalent to essential disorder. In Hegel's logic the contradiction between the two gives rise to an instability, an eternal passage from being to not being and not being to being. Such a process is not however mere oscillation; for the passage from being to not-being and back is *becoming*, the emergence of quality, the development of the Universe. In exactly the same way

195

the passage from order to disorder and back again is not a mere oscillation, but the generation of novelty. Each relation, because it gives rise to a new order—i.e. a new domain or hierarchy—also makes possible a new disorder, a new shuffling of that hierarchy. But this new disorder again makes possible a new disorder—a still more complex domain. Hence evolution consists of the development of domains.

Otherwise it is impossible to attach a meaning to the emergence of such a complex order as that of life. To regard a miracle of order such as the human body as more disorderly than the initial nebulosity we have postulated is plainly absurd, yet the law of Entropy as usually phrased demands just this. Indeed this patent contradiction has given use to the suggestion that life sidesteps the law of Entropy. There is no evidence for this and it is in fact unthinkable, but in any case the same criterion applies to crystalline structures. At the same time it is true that there is a constant reduction of availability of energy due to an increase in disorder.

In fact therefore becoming means that there is a continual breaking-up of domains or orders. This is the shuffling of the like quantities which as we have explained elsewhere is the basis of prediction, determinism, and necessary connection. But each breaking-up, or domain-death, involves the emergence of a 'higher' domain or order. This is the emergence of unlike qualities and the basis of novelty, causality and accident, quantity becomes quality.

How is it then that the law of Entropy could be formulated as a law of the increases of disorder, and the lessened availability of Energy?

Because it is a physical law. Physics is concerned with generalizations about quality as it becomes quantity. By definition as it were—because of its generality—physics is debarred from considering quality except under the aspect of quantity. It must not consider novelty and accident until it has already been gathered into oldness and necessity. It can only consider the like, the necessary connection, in all phenomena. Hence physics does not so much give rise to the law of Entropy as its province is defined by it. For by definition as it were all higher order is excluded from its province, into which enter only domains as they are broken down into disorder. It is the death science, because coming to life is excluded from its purview. Since therefore it is allowed to consider order passing into disorder, but not disorder passing into order, since it may consider quality becoming quantity, but not quantity becoming quality, it must necessarily find the law of Entropy true throughout its domain; for the law of Entropy defines the domain of physics. All order passing into disorder is a physical problem— nothing else is known to it. And since there is a continual passing of order into disorder as well as in the reverse direction, such a sifting will necessarily give us a perfectly unique serial world. Hence the predictive power of physics —because it deals only with likeness and necessity and appears to produce everything from a primal nebulosity with certainty. But it can only predict recognizable entities and in fact its prediction is always based on the prior breaking down of real concrete entities to their basis of nebulosity or utter generality Such a process may be called the recognition of physical entities and imports the accidental element into physics. Physics

197

talks about events with the arrogance of certainty, but we can never be sure that it is talking about the event we have experienced or expect to experience. Since for the purpose of physics each new entity or domain as it emerges has to be broken down to an abstract nebulosity, it is no wonder that the emergence of new entities, or qualities, which is Time, appears in physics as an increase of disorder. Every increase in complexity makes possible an increase in disorder—a well-furnished room can be more untidy than a monastery cell. Hence the disorder of Entropy is in a sense artificially created. This conception of Entropy is important as affording a bridge from the lifeless, hopeless world of physics to the rich pushful world of biology without falling into mechanism or Vitalism.

The same meaning must be attached to the assumption of the lessening availability of energy—for energy, as our analysis by now should have made clear, is the most generalized component of quality. Energy, the quantum, quantity, is the likeness in all quality.

Hence a primal nebulosity, a world without quality, is a world of all-available energy. But since quality can only exist as a relation, as a connection in which the unlike emerges, it is a world without available energy. Energy is a qualitative difference in particles reduced to it most similar components.

Thus the continual decease of available energy between particles is matched by an increase in the available energy between systems or domains. It is a process like the passage from the coarse all-or-none reactions of instinctive organisms to the fine integrated reactions of a man; one is more complex and selective than the other. Thus

side by side with the reduction of inter-particulate energy
levels, is the increase of inter-hierarchic energy levels.
This does not mean necessarily even that the levels are
fewer, for the domains can cut across each other, and the
number of domains of integers is higher than the series
of integers itself (Cp. Cantor). Eddington visualizes the
process as the gradually running down of a once fully-
wound Universe. The world starts to die directly it is
born. Apart from the contradictions inherent in the
conception of the 'start' of the Universe, this conception
is simply the result of the restriction of the field of
physics, necessary if the complexity of the Universe is
to be mastered for the purposes of prediction. By restrict-
ing itself to particulate qualitative differences, physics
restricts itself to a form of energy whose availability
must necessarily grow less for its non-availability in this
form is the condition for its appearance in more complex
forms of availability—as qualities of domains. Quantum
quality becoming domain quality, physical energy becoming higher energy (chemical, biological, or psychological)
is the part of reality's becoming which physics, for purposes of abstraction, completely deletes.

'Although the Law of Entropy by itself is not sufficient
to determine the direction in which the state of a material
system will change in the next instant, it excludes always
certain directions of change, the direction exactly
opposite to the one which always occurs being *always*
excluded.'

Of course this statement is tautologous. For a return
to a direction exactly opposite to the one which actually

199

occurs, is in fact a change to a state already left. But such a change would be indistinguishable and hence it would be impossible to say that it was different—i.e. that one had gone back.

But it is important to note that this law is only true for closed systems. But a really closed system is completely detached causally from the Universe and hence is unknowable. But if there is any connection with the Universe then one has a background against which to orient the change. There is then some meaning in the statement that the system has gone back to a former state because one can distinguish former from latter state (although in themselves identical) against the background of the Universe.

Of course there can be no closed system in that sense except the Universe as a whole; any closed system would be by definition unknowable and excluded from the Universe. Therefore the clause about always excluding the direction exactly opposite to what occurs, is simply a most general statement of the characteristics of Time. We have already explained this in a different form; there is a constant emergence of unlike from like and its ingathering as a homogeneous like into a new unlike. Taking any three states like (past) like + unlike (present) unlike (future) it is evident that it is impossible for unlike to be stripped off as a difference to produce the future, because it would be the past. Hence the law of Entropy is not 'intimately connected with the irreversibility of processes'; it states the irreversibility of processes. It explains that, taking the Universe as a whole, becoming has a certain universal characteristic which is what we mean by Time as immediately experienced by us in the passage of past, present and future. This universal characteristic is that the present can in no circumstance become the past. Time flows. Newness emerges. All is becoming. This is such an obvious statement to make that only the metaphysical mechanism of bourgeois physics, which excludes living concrete time, could make it seem so bizarre.

Other certainty: probability

Consequently that part of the law of Entropy which states that in a closed system a direction exactly opposite to the one which actually occurs is excluded always, is a general statement of the meaning of time in the Universe. We saw that an event can only be regarded as certain in all its aspects when it is in the past; hence a law which is virtually a specification of Time can contain an absolute certainty. However owing to its restriction to a 'closed system' it is only true as a generalization. The Universe as a whole cannot go back in time. This does not however prevent certain parts of the Universe from 'oscillating' against the general background. Such oscillations however because of the nature of the universal law, necessarily involve a corresponding change outside. For example, given a Universe divided into two parts, system and not-system, then any reduction in entropic disorder inside the system must be accompanied by an increase in entropic disorder outside, if the law is to be true for system + not-system as a whole, i.e. as a universal law. This provides a causal connection between the two of a thermodynamic kind, but this very connection has as it were produced a higher order, the separation of system from non-system, which did not exist before the emergence of the entropic disorder.

It is this which has led to a misunderstanding of the nature of entropic disorder. It is a disorder of the transactions of particles. As such it is true that 'nothing in the statistics of an assemblage can distinguish a direction of time when Entropy fails to distinguish one,' as Eddington phrases the Principle of Detailed Balancing (i.e. to every type of process (however minutely particularized) there is a converse process, and in thermodynamical equilibrium direct and converse processes occur with equal frequency).

Now if to every type of process there is a converse process and these processes are atomic, then it is obvious that a state of thermodynamical equilibrium can be reached. But in fact it is untrue that to every type of process there is a converse process, for by converse process is meant a process exactly similar to the prototype except that it is different in time—i.e. against the background of the rest of the Universe, for there is no abstract absolute time. But even so there must be some ingression of time into the process itself, for given a process ABC and repeated CBA while the background proceeds $A'B'C'D'E$ while it is taking place, then unless in the first process there is a relation $B'A$ starting the process and $D'C$ finishing it and in the second case relations CB' and AD' how can the two processes be distinguished one as prototype and the other as converse? But if they enter into the background in this way there is an ingression of new into them, and one is no longer the exact converse of the other.

But in another way, if 'converse process' is to have any meaning, it must be epistemologically distinguishable from the prototype—it must be different in time. There must therefore be a real newness in it; hence the conception of exactly opposite processes is a mere abstraction of limit. In fact all processes are different, but converse processes—i.e. processes which have in them the minimum ingression of quality from outside and are yet distinguishable—represent the least different.

They are in fact processes considered as already accomplished, so that the unlike in them, becoming like, they seem opposites which have lost their difference in unity. But they can only be known in a relation which involves a real ingression of newness.

Hence the Principle of Detailed Balancing is only true as an abstraction or limit. It postulates exactly converse processes and therefore removes the time out of process. It is correct therefore to say that in such a world no Time exists, for the processes have been sterilized beforehand; but in fact such a sterilization makes them epistemologically meaningless.

This limit however involves a definite conception, that of converse processes. If Being consists of order passing into disorder and *vice versa*, how can converse process be given a meaning—i.e. under what conditions will the former process exactly resemble the latter, so as to be robbed as far as possible of time? Only if the process of disorder passing into order is decomposed, so that it is split into its parts, and no longer therefore contains the new order which is a property of the whole. For example a man is born, a man dies. The one process is a converse of the others, and yet it is wholly different. The first is

203

synthesis, the second is analysis. If the two processes are however completely split up into their basis—the movements of atoms, the processes will be found to be similar. But this splitting up means the stripping of those qualities of order and difference which occurs in time.

Hence the law of Entropy is only true as a statistics of crowds. Only in so far as it splits all processes into processes of indistinguishable particles or units considered not as systems or domains, but as mere crowds, is it correct. Such units can only become more disorderly because the other type of order which emerges from this disorder is by definition excluded for it is an order of domains—i.e. of 'converse' processes which are not reversible. Because the law of Entropy reduces everything to its fundamental likeness, it is an abstraction and absolute Principles like that of Detailed Balancing are also abstractions.

This does not alter its penetrative power. Prediction depends on the persistence or ingathering of like into physics, by its remorseless analysis of all processes to the like basis, it gains its predictive power over Nature. Statistical laws such as that of Entropy are only one aspect of it. Organizations are reduced to crowds of units.

Molecules in life different—Dice reduced to like events fallacy Particles—Best but ultimate unlikeness—Every particle different—No time—Nebulosity—Probability laws—Other directions—Entropy and quality of organized crowds but one which excludes organizations (Disorder produces order) —Measure number of order as well—Eddington's view—Predeterminism—Practice knowledge of certainty

In fact this is incorrect. For every particle has a difference (not 'in-itself,' since the basis of particle abstraction is reduction to ultimate likeness) but as being part of a different system. It is not merely that a molecule in a living body is different from a molecule in a dead body; it is that any molecule has what may be called a domain difference to another molecule. Even in the simplest case of all—a particle in itself, it is still a known particle— i.e. a particle as part of a simple object-subject system. Hence by reducing processes to the statistics of units, thermodynamics, of which the law of Entropy is a part, reduces being to a crowd of particles-in-themselves. No such particles exist, for they would be unknowable. All particles have relations among themselves and it is these relations which constitute domains. They are however ruled out by statistical physics.

This for example is fairly clear when we take simple examples of statistics. They are all examples of particulate events. Probability is based on the assumption of a Universe of n discrete events. For example a chance of a die falling six upwards is 1 to 6. This calculus is based on the assumption of six possible throws, all equally likely, of which only one can be realized. But the equally likely means that there is no difference between them in the domain surveyed. But if there is no difference between them how can it be detected that a six and not a one is thrown. Hence in this form it is tautologous. If we have n events, indistinguishable from each other, of which only 1 can be realized, no meaning can be attached to the statement about the chances of any particular 'one' being realized. There is no particular 'one.' They form a continuous stream, and hence the probability calculus

$\frac{1}{n}$ is simply a dimension, a section in time by which $\frac{1}{n}$-th of a dimension n is discriminated.

Of course we can in practice discriminate between the throws because of the black marks. But the difference in the marking is a physical fact which either affects the evenness of the events, or if it does not, belongs to a domain outside the field and therefore reminds us that it is not an absolute field but only an abstraction. And it is impossible for differences outside the field not to have some effect on those inside—e.g. the difference in marking must have some microscopic effect on the chances.

If the events are indistinguishable (really equal chances) the probability would be certainly because the discrete events would have blended in one event of which the probability would have become a mere measure. Of sixty identical seconds the probability of one being the given second is no longer a probability—it is the certainty of a second being a sixtieth of a minute. And this certainty has become a meaningless certainty, it is simply a consistency of measurement. Hence probability is always the sign of an abstraction in which differences are disregarded.

We all know the belief that if the molecules of the die could be followed their motion could be established with certainty by an integral calculus. This is merely another way of saying that if we could break down sufficiently the events we could get to like particles which, by shaking off the differences that made the major events hazy, would give us a predictive certainty.

Eventually we get to the elementary particles, with which our statistics involve enormous figures. But now we hope to escape from statistics into certainty. We find

206

however as we have seen above, that the consummation
is prevented by the nature of discrete particles themselves,
whose velocity and position can never be precisely known.
But this in turn is because the knowing involves a
subject-object relation as a result of which the particle
ceases to exist-in-itself. If the particles were disposed in
a matrix of space-time then the conception of particles
in themselves would be correct, and particulate certainty
would be the basis of particulate probability. But space-
time is shed by the likeness and unlikeness in the flux of
becoming, and hence the standard particle is an abstraction. The Universe does not consist of n particles in
themselves but of n particles in relation, and the domains
formed by these relations ensure the difference of these
particles. And the simplest domain of all, the knowing
of a particle, involves a uniqueness.

Thus the particle-in-itself involves uncertainty, because
it is an abstraction. The use of larger and larger numbers
of particles involves greater probability, because the
difference is averaged out in wider and wider domains.
Partially closed systems produce greater certainty because
they are domains. The only absolute certainty is however
the whole Universe, because only then are all domains
included in the generalization without being cut across.
But it is precisely then that we understand why only the
past is certain—for the Universe includes the knowing
particle and the particle can only know, as object, its
past. It can never know the Universe including itself in
one present, and hence cannot acquire complete predict-
ability of the future.

Eddington appears to believe that Entropy is unique
because it measures the organization of a system; thus

207

it differs from analytical physics, which hopes to plot in detail the movements of particles. It is true that Entropy does not measure individual particles; on the contrary it measures not-individual particles—particles as units. And it is precisely because of this that it does not measure the organization of a system but the disorganization. It is the reverse side of a tapestry—all loose ends and confusion. The particles derive their individuality, not from any individuality-in-themselves, but from their forming part of domains. Individuals are intersections of species; and equally species are aggregations of members. Entropy is concerned with species as mere aggregations of members, and as such it does not measure the organization but the parallel disorganization. This gives it predictive and organizing power, because a crowd of like individuals persisting in time is a fundamental abstraction for prediction.

Thus we know how much reality to attribute to the Eddington picture of a Universe running steadily down hill; dying as soon as it was born, to a state of maximum disorder. It is a parallel to the Universe as it was born—fully wound up. They are creations resulting from the piling-up of abstractions at each end of the field. For the Universe running down is also the Universe evolving, and the dialectic generation of new domains and new complexities. The fully wound-up Universe is also purer Being—the Universe fully run down. If the kind of order (domains) which the Universe produces as Time is excluded from consideration, then nothing can be found in the Universe except the shuffling of units—i.e. the increase of disorder. But this abstract view, which excludes from reality real Time and real quality, in fact gives rise to contradictions which cannot be resolved without passing beyond science altogether.

Therefore when we say that physical laws are statistical laws, we merely define the domain of physics, which is that of likeness as it is ingathered. This reduction of particles to units excludes their small difference, and this difference may always add and emerge in the effect as a big difference, as accident. This accident is however an aspect of necessity and means that in spite of physics, as it were, a new domain has been generated. The greater the number of particles however the smaller the likelihood of a new domain emerging. But with single particles each event constitutes a unique elementary domain, hence its uncertainty is complete. But it is complete only within the limits of the domains which is of course (with the elementary domain) the quantum h and the quantum h is certain. Statistical probability does not therefore exclude determinism, it is an aspect of it. It is the form taken by the certain (i.e. unique) quantum h's when they are taken as all similar. Their only similarity is that of forming one Universe, which therefore again achieves certainty but at the price of including the knowing atom and becoming an unknown certainty. .

Hence we have the quantum and the Universe both of which are completely certain and completely uncertain, and where therefore neither term has any meaning except as a limit. Between are numerous particles. There is both certainty and uncertainty, one determining the other, but they are separable. In physics the uncertainty is separated off, the tapestry's reverse side is viewed, the particles are regarded as like units, and therefore we regard them under the aspect of probability. Alternately

P

the certainty can be separated off, the particles are regarded as individuals with predicates, i.e. as part of domains systems, and we regard them under the aspect of order.

And this should bring us to the final and yet deepest rooted problem of physics.

CHAPTER TWELVE

Causality and freewill—Rooted in bourgeois gen.
—Mechanism—Future

THE uncertainty which is enunciated in the Heisenberg
Principle has been hailed by bourgeois scientists such as
Eddington and Jeans as at last giving scope for freewill
in Nature and freeing man from the nightmare of
bourgeois determinism.

We can understand this cry of relief better if we revert
to the analysis we made at the beginning of the study as
to the concept of mechanism. Mechanism, as applied to
the object, inevitably leads to bourgeois predeterminism,
that is with the fact that all history must unroll itself with
iron precision from the initial case; that a calculator can
predict the whole of the future with precision.

We saw that this nightmare came about because in the
bourgeois economy Nature, the object, was gathered up
completely into one class of society as the machine, and
it seemed that the other class could control it completely
by a simple one-way property right which was given him
because, in the process of differentiation, consciousness
had fallen to him as a privilege. Hence consciousness,
linked to the property right which veiled a coercive
domination over men, seemed in itself to guarantee a
complete control over Nature of the one-way character
involved in ownership. Thus the object became the object
in contemplation, whose necessity was known merely in

211

theory. All qualities of the object which involved a mutual determinism of mind and object denied this lordly relation to Nature and hence were regarded as not really in the object. This gives rise to mechanism. Ultimately Nature is stripped of all newness, of all the unpredictable, and hence can appear as something whose future can be wholly and necessàrily determined by pure knowledge without practice. This is all predeterminism is. Moreover the purely contemplative relation of the bourgeois to Nature as seen in the machine, in the division of labour which is Nature practically ingathered into society, but here only one class of society, makes him believe that the Universe as a whole can be contemplated, in spite of the fact that the knower forms part of it.

Meanwhile all the newness and subjectivity which is really a mutually determining relation between Man and Nature, having been cut loose of Nature, floats about freely and seems to be purely human and 'spontaneous.' It is developed quite independently of Nature. Thus, since Man's relation to Nature is supposed to be one-way, like a property right (Man commanding Nature as the machine, but the machine not determining Man), all these newnesses and subjectivity seem to be determined by nothing and just emerge as pure chance. Thus all the specifically social qualities, generated by the struggle of Man with Nature, which constitute the 'freedom' in Man's relation with Nature, seem to leap into being without cause, because they have been already stripped from the object. Freedom seems to lie in subjectivity as it emerges in bourgeois society, that is, with its causes concealed. In bourgeois society *freedom is the ignorance of necessity.*

This in itself is only another aspect of the condition of bourgeois production which is commodity production in its most developed form. And in commodity production Man has lost control of his social relationships. His desires appear out of the market blindly, without visible cause. They wax and wane like an unknown force of Nature. His goods disappear into the market, and no one knows whether too much or too little has been produced; whether his production is a social good or a social disaster. Yet this is the 'free' market for which bourgeois society fought everywhere. Hence the market is the kind of chasm between Man and Nature, producer and consumer, bourgeois-class and machine-class which seems to be a hiatus but which is in fact merely the penumbra of the bourgeois ignorance hiding the causal active network of relations between the two. Thus all the newness and causality in Nature are separated from all the likeness and determinism and so abstracted the two appear as irreconcilable—as subjectivism and freedom on the one hand and objectivity and determinism on the other. Teleology and purposiveness is a reflection of mechanism. Man makes a plan, the plan is carried out by the machine; one is necessary to the other. But the machine is not completely determined by the plan; previous experience with machines determines the plan; they are the subject of a dialectic evolution. The machine, Man's interpenetration with Nature in practice, generates the plan, Nature's interpenetration of Man in theory, and *vice versa*. But once the machine and plan are placed in a one-way relation by mechanism, then teleology, in which the plan appears as 'spontaneous'—i.e. implanted by a higher Mind—must necessarily emerge as the philosophy of subjective qualities.

Yet although bourgeois economy separated subjective freedom from objective necessity, the separation can never be completely made. The bourgeois body is both object (in others) and subject (in oneself) and hence only in solipism is subjectivity free. But even here there is no separation; objectivity is simply denied, and this makes mechanism, with its important control over Nature, completely meaningless; hence its unacceptability.

But otherwise the body as body is an object; it is subject to necessity. The body as mind is a subject; it is free. But what avail for the mind to be free if all the body's actions are predetermined? This is the bourgeois nightmare from which Jeans and Eddington hope to have escaped by the following solution; even the object, after all, is not subject to necessity, but only to probability. In the light of our previous analysis it will be easy to detect how much freedom there is in the bourgeois sense, in the object.

But first it must be pointed out that the older scientists, such as Einstein and Planck, do not believe in this abandonment of mechanism and determinism. Einstein is particularly contemptuous, as might be expected from the fact that his system represents the climax of the mechanistic scheme. The following dialogue took place:

'EINSTEIN: "Honestly I cannot understand what people mean when they talk about the freedom of the human will. I have a feeling, for instance, that I will something or other; but what relation this has with freedom I cannot understand at all. I feel that I will to light my pipe and I do it; but how can I connect this up with the idea of freedom? . . ."

'MURPHY: "It is now the fashion in physical science to attribute something like freewill even to the routine processes of organic nature."

'EINSTEIN: "That nonsense is not merely nonsense, it is objectionable nonsense."

'MURPHY: "Well, of course, the scientists give it the name of indeterminism."

'EINSTEIN: "Indeterminism is quite an illogical concept." '

In fact Einstein here gives an excellent definition of freedom. But since he has created a completely mechanisitic world, there is no room for freedom in it, and he cannot understand what relation freedom has to such a Universe. In fact it has none, for it is excluded by the categories of mechanics.

Planck adopts a slightly different attitude towards the question of freewill.

'The existence of strict causality implies that the actions, the mental processes, and especially the will of every individual are completely determined at any given moment by the state of his mind, taken as a whole, in the previous moment, and by any influence acting upon him coming from the external world. We have no reason whatever for doubting the truth of this assertion. But the question of freewill is not concerned with the question whether there is such a definite connection, but whether the person in question is aware of this connection. This, and this alone, determines whether a person can or cannot feel free. If a man were able to forecast his own future solely on the ground of causality, then and then only we would have to deny this consciousness of freedom of the

215

will. Such a contingency is, however, impossible, since it contains a logical contradiction. Complete knowledge implies that the object apprehended is not altered by any events taking place in the knowing subject; and if subject and object are identical, this assumption does not apply. To put it more concretely, the knowledge of any motive or of any activity of will is an inner experience, from which a fresh motive may spring; consequently such an awareness increases the number of possible motives. But as soon as this is recognized, the recognition brings about a fresh act of awareness, which in its turn can generate yet another activity of the will. In this way the chain proceeds, without its ever being possible to reach a motive, which is definitely decisive for any future action; in other words, to reach an awareness which is not in its turn the occasion of a fresh act of will. When we look back upon a finished action, which we can contemplate as a whole, the case is completely different. Here knowledge no longer influences will, and hence a strictly causal consideration of motives and will is possible, at least in theory. If these considerations appear unintelligible—if it is thought that a mind could completely grasp the causes of its present state, provided it were intelligent enough—then such an argument is akin to saying that a giant who is big enough to look down on everybody else should be able to look down on himself as well.'

By strict causality, Planck here means strict determinism, and in fact his definition implies the non-existence of the causal relation, for knowing becomes a mere awareness, a mere inner activity following outwards; it is not a causal subject-object relation.

216

He bases the freedom of the will therefore not on this relation in all its fullness but in the impossibility of the subject knowing itself as object immediately. We have already dealt with the impossibility of knowing the Universe as a whole. But Planck makes this the basis of the freedom of the will whereas it is in fact the reverse— it is the limitation to the freedom of the will ever becoming absolute. It is a barrier to freedom, not a source of it. Planck reasons as follows: A man would be unfree if he experienced the causal relation in which he is involved; as subject he cannot do so; therefore he is free. Implicit in this reasoning is the bourgeois premise; freedom is the ignorance of necessity. In determinism the causal relation is regarded as the object—the activity of subject and object becomes objective to another'subject. For this to take place the relation must already be part of the past, and the new activity have become congealed into the extant sum of like. Hence freedom has gone out of it. But this does not mean that because Man cannot experience his causal relation to the Universe objectively he is free. If that is so, a stone has more free will than Man, because it certainly cannot know its causal relations objectively. By this definition with its implied ignorance of necessity as freedom, Man is the least free entity in Nature. Man is free because and in so far as he can experience his causal relation to the Universe not objectively, but in its fullness, in a higher richness, as *knowing* subject. Freedom then is the consciousness of determinism. This is the opposite of the bourgeois definition.

But Man can only become conscious of determinism by his consciousness of a causal relation. And this at once makes consciousness a unity with practice, with action

217

upon Nature. Consciousness as freedom is derived from Man's practice on objects. He wills to light his pipe, and he does so—i.e. the lighting of his pipe actually occurs. He lighted the pipe because he willed to. Only the complete relation constitutes freedom. For example if he willed to light his pipe and did not do so (because he was bound or paralysed) there has been no activity, and the effect—the lit pipe, has not appeared. This is not a genuine causal relation. He is unfree.

Or again suppose he willed and believed he lit his pipe, but no lit pipe appeared in the social field of vision, we should say he was deluded, and would not regard him as free, but as a man 'compelled,' as a sufferer from hallucinations. Hence the effect, the practice, which completes consciousness and makes it freedom, must be real practice and not a mere personal belief. It is practice socially recognized by society, which codifies it and gives it a conceptual clothing.

Yet this does not mean that practice is freedom. It is the consciousness in its fullness which changes it and makes it freedom. A man sleepwalking or the autonomic nervous system's activities are not free because they are unconscious; and we can perceive a difference between free and unfree activities. They belong to a new domain. Hence freedom is the consciousness of causal relations. It is a conscious of activity in the world and the emergence of the new, not as mere activity, but as a complete relation containing necessity.

For example, the division of labour, machinery, knowledge of the seasons, all give men freedom in relation to Nature. Freedom is control, power, efficacy. But it is precisely because in the division of labour, freedom, knowledge of the seasons, that man is conscious of the necessity of nature, of the strict determinism or reign of law in it, that he is free of it. This is a social freedom, a freedom of society in relation to Nature, and it is embodied in a social consciousness, science and technology, which is inextricably united to a social practice. Man is free of Nature by and through changing it. Man's freedom in relation to Nature is economic production.

But Man is not yet free in himself. Society is still ravaged by wars, slumps, neurosis, unemployment, civil war, and poverty, because Man is not yet conscious of the necessity of society, of the inner laws of society. That is to say, he is not yet conscious of his own necessity, for it is only as object, as 'other men,' that he can know himself objectively. This however is what we should expect, for bourgeois man interprets subjectivity, Man's freedom in relation to himself and his desires, as the *ignorance of necessity*. This is however the definition of unfreedom. And we saw that this definition was generated by the darkness of the 'free' market, which everywhere stood between men and Nature, between subject and object, and seemed a chasm, not a mere unconsciousness. And this free market itself is only a generalization of the anarchy of commodity production, that no conscious relation unites producers and consumers, but all works blindly. Thus society is ravaged by the market as by a blind force of Nature, just because society is unconscious of its necessity, its inner laws. Its inner law, the road to freedom if known, is a catastrophic compulsion while Man is ignorant of it. And the market itself, the chaos from which it sprung, is a product of the special property right which divides the bourgeois owner from

219

the 'Nature' he owns, which is in fact not Nature, but Nature as the machine, the nucleus of a whole proletarian and exploited class. The bourgeois nightmare and dilemma regarding freedom is rooted in his class society and the separation between theory and practice which it begets.

Self-control is evidenced by control of the sum of selves which make up society—by revolution.

And freedom in personal consciousness is the same recognition of necessity. If I am suddenly propelled from behind, my movement is unfree. If a tic or St. Vitus's dance propels me forward I am unfree. For in both cases although aware of the movement I am not conscious of the causal relation as a subject. The new quality which is part of the causal movement does not unfold itself in me as a subject. Only if the action is willed—i.e. if I am conscious of the motive—am I free. This volition is one aspect of the new quality as it emerges—it is the heart of activity. Activity wherever it is seen as an object (as when I objectively regard my body as being pushed) appears as the causal relation, but it is then already part of the past and its quality has become quantity; it has fallen into the province of certainty. It has become determined; it is the causal relation, theatre of power and activity, as it sinks into determinism. It passes away, not completely, but into the ground of a new quality. But activity as freedom is the causal relation as experienced by the subject, and this consciousness already lifts it into a new domain. And this freedom is inseparable from the passage into the effect—i.e. into practice. Change must be produced in the object as a sign of efficacy and part of the relation. The man cased in plaster of paris is not free, although his will is free. He may perhaps be free in this much, that he can 'change his mind' and thus change in the body; an object produced by the body as subject (one part set against the other) is the sign of practice.

Not all the causal relation becomes conscious. Freud tells us of 'motives' in the unconscious which sway the will. To that degree a man's will is unfree; he is compelled. The road to freedom consists in making those motives conscious. Once again freedom is the consciousness of cause. But this is true also of outside causes which unconsciously affect man's will.

Part of the bourgeois error is rooted in his misunderstanding of the social genesis of freedom. Man's knowledge of the necessity of Nature, and his actual change of her, are tasks only possible in any fullness to associated men. Practically all Man's knowledge of Nature's structure and practially all his change of that structure is socially achieved.

But this in itself imposes necessary laws upon men which are social laws. They appear as the division of labour and the organization of it. These social necessities are evidently the means by which freedom is generated for Man. It is not an abstract entity but a real activity whose proportion can increase.

But to the bourgeois all these social relations whereby freedom is produced appear as restraints, because historically the feudal relations of society at a certain stage had to be destroyed to permit the further development of economic production—i.e. the raising of freedom to a higher plane. The bourgeois slogan became, 'Away with all social relations that limit our freedom.'

221

But the further generation of freedom, i.e. the further development of economic production, by producing a further division of labour and a further organization of it, involved an increase in complexity of social relations. But since they were not acknowledged by the bourgeoisie, and in fact came to stand in contradiction to the only relation he acknowledged—individual ownership of the means of production—these social relations were veiled and hidden beneath the overt structure of capitalist ownership. The real relations of men round machines in factories was plastered over by share certifications and relations between employees, directors, and investors. The denial or unconsciousness of these real relations whereby freedom was generated took the form of the free market; such an unconsciousness is the character- istic, the definition of commodity production and is precisely why in such an economy 'man has lost control of his social relation.' Hence the assertion of individual liberty from social restraints as true freedom and the denial of social relations as restraints is inextricably part of the bourgeois world-view. Man is ignorant of the motives of his thought and his behaviour and his con- sciousness and hence loses control of society, and himself as one of society's selves. A correct conception of freedom can only be generated by that class whose organization already reflects the division of labour, the object Nature, which it has sucked into itself to transform, and which causes it to expand to include all society, including the consciousness which has broken loose from it and become confused. To such a class freedom is the consciousness of necessity but its successful struggle can only be accomplished by a recognition of those relations and causes which make society what it is, and have the potentiality of making society what it can be.

Against this background it is possible to understand why scientists such as Eddington, Jeans and Schrödinger welcome the newer developments of quantum physics as a charter of bourgeois freewill.

Their situation is roughly this. Freedom consists in spontaneity; an apparently uncaused emergence of quality. How can this be reconciled, in material human creatures, acting in a material world, with strict determinism—a necessary and predictable connection between all events, such that all the future can be logically unrolled from any given state?

We are able to see that this is an imaginary dilemma, which only arises because bourgeois theory cannot see freedom as an aspect of determinism, or accident as an aspect of necessity. Instead each is abstracted from the other and the scarred, torn away side forms an impassable barrier between the two. Determinism and necessity become crystalline and incapable of evolution. Freedom and accident float about without a root.

Heisenberg enunciates a principle of uncertainty and Thermodynamics a universally applicable law of statistics. Schrödinger formulates atomic science in terms of probability waves. Eddington and Jeans at once assume that this is bourgeois spontaneity, an absolutely uncaused effect. For example of 500 atoms 'exactly similar' 400 may go into state 1 and 100 into state 2 and according to Eddington, there may be nothing in the atom to distinguish which is destined for one state and which for the other. They are all identical. We have already seen that it is impossible to talk of 500 identical atoms; they are

223

indistinguishable from 1 atom encountered at different times. Therefore they are all different. But their differences fall into domains which emerge as State 1 and State 2. It is true that we cannot predict this; but it is precisely domains that we cannot predict, they emerge as order by the increase of disorder (i.e. atoms all grouped in one State falling into 2 States). Therefore before the emergence of the domain we can only speak of probability; after its emergence there is a necessary connection between the various differences which group themselves into State 1 and those that fall into State 2. We can now speak of certainty. When Eddington says there is no difference between them such as to cause the falling into two different States, he is being misled by his own abstractions. Evolution consists of the generation of differences or domains which are however real sortings or groupings. Once divided, the division has a real basis.

This is completely different from the idea that Nature has bourgeois freewill; i.e. that anything might happen. The bourgeois freewill does not permit of statistical laws. For though there is a probability of say $\frac{1}{1,000}$, the probability itself is certain. Hence this would show there is a rigid limit to bourgeois freewill.

In fact of course once one leaves the abstract conception of freewill as it emerges in bourgeois philosophy, and regards the real experience, it can be seen that there is no relation between the real experience and probability. Yet it is precisely to intuition that this philosophy always appeals for its conception of freedom as the consciousness of necessity. For example, if having willed to raise my arm I felt there was only a probability of doing so, I should not regard that uncertainty as freedom.

On the contrary, I should regard it as a limitation upon it. Again, since according to the laws of Thermodynamics it is always possible that the kettle placed on the fire may freeze, it is always possible that my arm may suddenly rise above my head. This also I should regard as an infringement of my freedom.

In fact therefore the reduction of all macroscopic laws to probability laws would be a limitation on my intuition of freedom. This itself shows that the bourgeois conception is an inversion of the immediate intuition. The immediate intuition of freedom is based on a strict causal relation—volition and action are necessarily connected. If they are only probably connected, it is a limitation of my freedom.

But even if the bourgeois conception of freedom were correct, the probability laws would not give man a very large freedom, even if the probability laws were strictly correct. We have seen that their form is due to an abstraction, that because they do not deal with the sum of particles in the Universe, they must necessarily exclude certain related effects, and therefore can only take the form of probability until they have actually occurred and become part of the Universe as a whole in concreteness. But even suppose they were absolute probabilities; the bourgeois conception of freedom rests on the denial of causality; hence human freedom would depend on the infringement of the causal relation which in this case takes the form of the major probability. B follows A takes the form of B may follow A. On these cases where it does not, the causal relation is infringed and there is the occurrence of freedom. With a body containing so many atoms as the human body, the chance that it will

225   Q

not obey all the old universal laws determining its motion is an astronomical number; hence Man must reconcile himself, according to this bourgeois philosophy, to attaining freedom once in several billion actions; thus there are probably odds against any human being yet born having performed a free action. This is certainly contrary to intuition.

It is not true dialectics to regard action as absolutely free. But then neither does intuition. Intuition feels that some actions are more purely free than others, but that all have some element of the unknown steering them, which represents a measure of unfreedom. But to bourgeois conceptions precisely this unknown force, this motive which does not enter consciousness is their freedom.

Eddington sees, I think, the fallacy in basing freedom on probability, but does not therefore give up his conception of freedom; indeed he could not do so without ceasing to be bourgeois. He therefore turns to the single atom. According to him, one cannot say beforehand whether an atom will make itself enter State 1 or State 2. We phrase this thus; we cannot say until the emergence of the new domain whether the particular uniqueness of the atom qualifies it for system 1 or system 2. However Eddington prefers the-in-my-opinion-less logical phraseology. He thus regards the atom's behaviour as spontaneous, or uncaused—i.e. free.

Now if it is uncaused, how regard it as probable that it must choose State 1 or State 2? And that there are quite definite and rigid odds as to how many atoms falling in its domain choose one and how many choose the other. Evidently if someone said to 500 individuals, 'you are free because 400 of you must be in Timbuctoo tomorrow and 100 in Baghdad and none of you can say which has one destination and which the other,' they would hotly deny it. For they are unconscious of the causal relation —they do not know the cause as subject in an efficacious subject-object relation which involves their desiring to go to Timbuctoo for definite reasons (or causes) and choosing it.

But even suppose the atom were as Eddington and Jeans evidently hope it is, so that someone could go to a crowd of people—'no one on earth can say where you will be to-morrow'; is this freedom? Plainly it isn't; for if anyone were to say, 'I shall go to Timbuctoo to-morrow,' he would regard himself as free only if he were able to turn up there 'accidents excepted.' And he would regard the accidents as infringements on his freedom. But with true spontaneity he might be anywhere and owing to the number of possible destinations the chances of his turning up at Timbuctoo would be astronomical. He would not regard this as freedom. Indeed it is doubtful if one can talk about chances against, for this assumes equal probability for all places; but it is doubtful if true spontaneity can permit 'equal chances,' since this involves a veiled causality and the emergence of a domain. In fact it is doubtful if true spontaneity has any real meaning. For it implies an event without a causal connection with the rest of events, and how can such an event be 'known'?

However to resume our argument. Suppose Eddington is correct and atoms possess true spontaneity. How does this permit human 'freedom'? Eddington's argument is that the whole body is a machine so balanced that the action of a single atom can 'tip the scale' between one

227

action and another; hence I am correct if I feel that at
one instant I can choose either of two given actions, as
I intuitively feel I can. But this argument is illegitimate.
Either the laws of thermodynamics apply universally,
or they do not. According to the laws of thermodynamics
the 'uncertainty' (which Eddington thinks is true
spontaneity) at atoms smooths out with numbers and
gives rise to probability laws. But Eddington wants to
assume that in human bodies they do not smooth out.
In that case the laws of thermodynamics do not apply
to human bodies—but certainly science has no room for
such a belief. And Eddington himself would be the last
to suggest that a body whose statistics are so astronomic
as the human body would disobey the laws of thermodynamics. This is like expecting kettles to freeze if placed
on the fire. He has fallen victim to a logical fallacy of
this kind. It is a law that Man may only expect a life of
thirty years, but individual lives are uncertain; therefore
I may be immortal. In fact however the uncertainty of
human lives is limited by certain states (0—110 years)
just as 500 atoms are limited to States 1 and 2. This
limitation is certain, and it is upon this certainty, and this
alone, that it is possible to erect a probability law. If
human life were really uncertain in the Eddingtonian
sense, it would be impossible to derive an expectation
of life. Thus we see probability is an aspect of necessity
or merely indicates the emergence of domains.

However even if Eddington's extraordinary machine
were possible—a balanced system ignoring the laws of
thermodynamics because a truly spontaneous atom was
at the helm, how could such a system obey the spontaneous atom if each atom composing the system was also

spontaneous? For if the machine is to work, the system must definitely respond to the atomic helmsman; that means it must be a causal determined system. Now either this causality is a probability law, and one of a new kind (for by definition the human body as a mass has become exempted from ordinary statistical laws) in which case Man's freedom depends on the helmsman's action being answered and therefore the nearer certainty the probability, the more Man is free. Any misfire of the helmsman due to the popping up of the hundredth chance, limits Man's freedom, which is opposite to Eddington's claim of uncertainty having a loophole for freedom, or else it is a strictly deterministic law, in which case Man is 'really free' in the Eddingtonian sense, but is only so because a new law has crept into Nature, thus defying Eddington's claim for Nature's universal uncertainty. This analysis alone exposes the contradictoriness in Eddington's conception.

But let us carry the analysis still further. Suppose the deterministically working system controlled by the truly spontaneous atom is in fact realized; does this correspond with the intuition of freedom? No, for it would mean that at any instant a man's behaviour might be anything; nothing in the world, including himself, could say with certainty what it would be. He might go to sleep, dance, have a fit, grin, gabble incoherently or discover the General Principle of Relativity. Now so far from corresponding to our idea of freedom, this corresponds to the behaviour of madmen—and we regard them legally and morally as unfree, irresponsible men——men who are compelled. But of course the compulsive causes are not non-existent but 'unconscious.' Human freedom intuitively demands that we can form certain volitions and fulfil them, any failures being regarded as limitations on our freedom. No such certainties, however limited, are possible with our truly spontaneous atom.

Finally if Eddington is correct, all objects are as free as human beings; the more atomic the object, the freer it is. This is contrary to man's intuition.

Hence we do not deny the bourgeois' claim to intuition. We believe that the final criterion of freedom is man's intuition engaged in efficacious—that is, free action—freely willed. But we deny that bourgeois conceptions of freedom are rooted in intuition. They are derived from categories which are simply the reflection of the sundering of theory from practice in bourgeois economy.

What is in fact the solution to Eddington's emphatic claim that man is correct in his intuition that at any moment he is free to choose whether he lifts his left hand or his right?

First of all, it must be pointed out that in regarding this as an important case of freedom, Eddington is misled by comparison with the atom which is 'free' to choose State 1 or 2. This is not freedom, but a limitation on its choice. (Timbuctoo or Baghdad.) The truly free atom could choose from a vast number of States. Similarly it is not freedom to be able to lift right or left hands only, but a limitation upon freedom. Of course Man is free to perform many other actions—the range of his causality is the measure of his freedom.

But faced with the decision to choose either action is he free to choose, as he feels he is? The answer is that he is free to choose. But he can only choose 'one' (or if he chooses to lift both, it is a third case). But having chosen one action, it is determined by his volition (which in turn has prior causes). If it is not, then the action is unvoluntary, and therefore unfree. Both actions are caused, but in one Man is conscious as subject of the causal connection and hence his action is free. In the other he is unconscious or (if he knows the cause—e.g. a push of his arm)—he is conscious, not as subject immediately experiencing the causal action, but of the complete subject-object relation seen as something objectively completed outside him.

Eddington sees the two actions like the two bundles of hay between which the donkey was equally placed. Would it therefore starve, or if not, which would it choose?

And this is only a recurrence of the completely identical atoms. There are no completely identical bundles of hay, except for purposes of abstraction; the only identity can be a commonness in the domain chosen. And the differences that make each distinguishable are, when the bundle of hay or action is chosen, the qualification for the new domain which emerges, the consumed bundle or the lifted hand.

In fairness to Eddington, he does not visualize the balanced machine and the spontaneous atom, as we have put it. Indeed he repudiates in one place the dependence of a decision on one key-atom. But in fact his argument here is so confused that it is legitimate to put it forward in this way, for his argument only has coherence, if this concept is adhered to. It is the basis of his whole argument. He only qualifies it in this way; freedom depends on the ability of mind to control the key-atom, and it is probably not merely key-atoms, but 'we must attribute to the mind power not only to decide the behaviour of

231

atoms individually but to affect systematically large groups—in fact to tamper with the odds on atomic behaviour.'

Eddington here introduces a new conception which completely negates his claim that it was possible in the year 1926 to believe in human freedom. For now the atom's behaviour is not uncertain, it is determined by another entity, apparently separate from the atom—the mind. So after all the atom is not spontaneous; whole groups can be affected in this way—it is possible to tamper with the odds on atomic behaviour—in other words with the second Law of Thermodynamics which in another place Eddington hails as the most certain of scientific laws.

So Eddington's argument eventually becomes: science reveals an uncertainty in atomic behaviour and the fact that all laws are statistical. But it is possible for the mind to make the atom's behaviour certain and infringe statistical laws; hence Man is free. Thus so far from modern science, as Eddington claims, supporting bourgeois freewill, this concept demands their abrogation.

The uncertainty of the 'individual atom' is based on this misunderstanding. The atom in itself is in its behaviour, object of no 'prior cause.' Naturally so, for if it were it would not be an 'atom in itself.' But in fact the 'atom in itself' is unknowable and does not therefore exist. It is simply an abstraction of thought. Taken as absolute, it gives rise to antinomies.

Jeans, Eddington, and even Schrödinger, all share this desire to prove that modern science permits freewill.

There is no stratagem too mean for this unconscious bourgeois illusion to use to buttress itself, simply because it is unconscious. Scientists, in their own sphere logical, will stoop to the most puerile argument to keep Man out of the causal world of physics. Jeans says, 'Nature no more models her behaviour on the muscles and sinews of our bodies than on the desires and caprices of our mind.' So to modern scientists not only our minds but even our material physical bodies, muscles and sinews, are not a part of 'Nature.' Human flesh is immune from causality and force is therefore an 'anthropomorphic' concept. In fact of course not muscular force, but our immediate experience of the causal relation—willing and producing effects in Nature by muscular action—enables us to generalize and discover change and causality and interconnectedness everywhere in Nature. Jeans also makes a distinction between 'subjective' and 'objective' probability. Objective probability is when 'even nature herself does not know the result of the experiment until it afterwards happens!' It is difficult to see what significance can be attached to this definition. How does Nature 'know' when something is going to happen? And how do we know she does not know? And does she know it is probable? Or do we only know it is probable? And if so, is not that subjective probability?

Again Jeans says that when birds fly through the air, 'their shadows on the ground beneath obey no uniform or deterministic laws, even though the actual flights of the birds may do so.' This remarkable discovery of Jeans ought to have been supported by a few instances of birds' shadows not obeying uniform laws, for certainly such a discovery would revolutionize science. For example, when did he find birds' shadows without a source of light, and without a bird interposed between

233

the shadow and the source of light? On how many occasions, and on what dates, did the shadows differ from the shape which one would predict they must have from the projective qualities of the bird's shape, the position of bird, source of light, and surface of the ground? How often, and to what extent, was the shadow's motion not determined by the movement of the birds in relation to source of light and ground? Obviously Jeans would not make such an inept comparison if he could discover a genuine case of complete accident or non-determinism in a relation. But it can only appear as an aspect of the determinism in a relation. And every relation must have an element of necessity in it or it is not a relation.

Jeans also conjectures and in some way connects it with indeterminism in Nature that the operation of life over-rides the second law of thermodynamics, and reverses the Entropy gradient. This suggestion was made in the nineteenth century and is now rejected by the unanimous verdict of competent biologists. But it does show that so far from modern science upsetting causality, as is constantly suggested, modern scientists can only introduce causality into the world by supposing an abrogation of the very laws they have discovered. And their readiness to do this is the truly disturbing feature of modern science.

This readiness to deny the findings of science is therefore not—as is often urged—the outcome of their researches. It is in opposition to their researches, and the following kind of argument is opposed to the spirit of science:

'The casting aside of all models and the wholesale

employment of mathematical formulas in their stead, because the latter are found more suitable for the representation of what is called ultimate physical reality, come very close to the Berkeleian standpoint and, in the theory of wave mechanics, reduce the last building stones of the Universe to something like a spiritual throb that comes as near as possible to our concept of pure thought.'

For what after all are these much despised models? They are bridges between the mathematical formula and the flux of nature. And the demand to do away with them (rather than to subtilize and purify them) is a demand to cut the bridge between thought and matter, between mind and Nature. It is a demand to cut the experimentation from physics. Models represent the interpenetration of practice (outer reality) and theory (mind) in the human consciousness. To do away with them veils the demand to do away with practice. And such completely negates science. Science without practice, without the appeal to experiment, is pure scholasticism and Alexandrian futility.

'This concept of the Universe as a world of pure thought,' says Sir James Jeans in his book *The Mysterious Universe*, 'throws a new light on many of the situations we have encountered in our survey of modern physics. We can now see how the ether, in which all the events of the Universe take place, could reduce to a mathematical abstraction and become as abstract and as mathematical as the parallels of latitude and the meridians of longitude. We can also see why energy, the fundamental entity of the universe, had again to be

235

treated as a mathematical abstraction—the constant of integration of a differential equation.

'The same concept implies of course that the final truth about a phenomenon resides in the mathematical description of it; so long as there is no imperfection in this, our knowledge of the phenomenon is complete. We go beyond the mathematical formula at our own risk; we may find a model or picture which helps us to understand it, but we have no right to expect this, and our failure to find such a model or picture need not indicate that either our reasoning or our knowledge is at fault. The making of models or pictures to explain mathematical formulae and the phenomena they describe, is not a step towards, but a step away from, reality; it is like making graven images of a spirit.'

If reality is a spirit—i.e. pure mind, then models are graven images of it. But most of us suppose graven images and models are real, and we are not insulting reality by making images of it. We also suppose that mathematical formulae are not material, and that we are insulting matter by reducing matter to nothing but 'mathematical models' of it.

But evidently if reality is pure mind, if a phenomenon is completely described in its mathematical formulae, practice is so much waste of time. Not experiment, but logic is the kernel of science. Once we have made a consistent mathematical picture of the Universe, we have learned all we can of it.

But of course this is contrary to the method of science. The test of formulae of reality is not their consistency, but their predictive power as proved in practice. If they fail their consistency becomes the very measure of their unreality, and a new frame has to be devised to fit the practice. It is just because the method of science is based on the assumption of a difference, a contradiction, between objective and subjective reality which can only be resolved by practice in which mind appears, not as a detached, spontaneous quality, but as part of a real causal relation—that logical consistency is not the test of reality. Science is enriched by the continual contradiction between models and theory on the one hand and models and reality on the other. The model is in fact the framework of an experiment. The model is a graven image or mimicry of reality; but so is the experimental set-up of a mock world. We learn about the world by changing it, and a change which is inconsistent with our self-consistent formula is precisely what extends our knowledge or reality. But the acceptance of Jeans' position in its fullness would mean that, like the Aristotelian Professor of Padua, we should not accept the inconsistency because in logical consistence, not in models or experimental observations, resides the real reality.

Thus the mentalism and tendency towards anti-scientific scholasticism of modern physicists is not a result of their researches, for it denies the method by which they were achieved. It is the result of a deep distress in bourgeois philosophy; of the contradictions between the world-view as a whole and their researches, and the mentalism and subjectivity springs not from their researches but from the decomposition of the world-view to which, in spite of its decomposition, they vainly cling. They turn their backs on their researches, and try to piece together the shattered fragment, with what results we have seen. The crisis in physics is a deep-rooted one —it is part of a general and final crisis.

The division of labour rises because by economic production Nature, the object, is taken into society, and society is differentiated and split up into specializations which correspond to the necessities of the object. The division of labour gives rise to real knowledge about the object resulting from real practice. In bourgeois society the division of labour is carried to its apogee, hence it gives rise to a unique knowledge of Nature.

But this knowledge is divided. It is specialized into departments, disciplines of trades. Hence the simple savage view of Nature as a vital glowing whole, everywhere developing like a real sensuous organism, gives place to a view of Nature as composed of objects and 'domains' (biology, ethers, physics, history). Nature is reduced to separate parts, each part different, and so reflecting a richer knowledge, but each part divided and so reflecting the division of labour.

But this division of labour is organized; the parts are subject to the integration of the work as a whole. The division is only possible because of the integration, which itself springs from the nature of the task. Hence this integration represents a body of information about Nature, which is a systemalization of existing information, over and above the enrichment due to specialization.

But in bourgeois society the division of labour is achieved as the summit of commodity production. And it is the characteristic of commodity production, that in it the division of labour is conscious (factories, trades, etc., consciously organized) but the integration is unconscious. The organization is achieved by 'the market' which is simply the name for the social unconsciousness of the organization. The organization asserts itself as a law, but unconsciously and therefore blindly, like a force of Nature.

Hence bourgeois theory can only see Nature as a sum of parts, of disciplines or fields, never as integrated, as organized. Each increase in division of labour makes that organization still less necessary, still more invisible. The wood grows more indiscernible for trees.

But in addition to this characteristic, commodity production produces a characteristic division of society —a class division. This gives rise to the aggregation of consciousness at one pole and practice at the other, as the result of a specific one-way relation (ownership of private property). This relation (the object controlled by pure contemplation) and this appearance of consciousness (freedom as ignorance of determinism) seems to bourgeois society the real integration of the division of labour. It seems to bourgeois society as if these categories are the real form and organizing principle of all the intricate specialization and division of labour which takes place in society.

In fact this is a fallacy. But it only appears as a fallacy, when the organization due to the division of labour bursts the organization imposed upon it by the specific relations of a class society,—the private property relation. The productive forces burst the productive relations. The real organization inherent in division of labour explodes the shallow and limiting coercive organization of capitalism. And with it explodes the bourgeois world-view—a shallow mechanical integration

239

based on just that real and rich knowledge produced by specialization. The bourgeois attempts to hold together the world-view by forcibly denying the very truths which threaten to explode it—and him—by denying objectivity and clinging to mentalism.

The new organization, once it emerges as the *conscious* (that is, controlled) organization of society, generates a new view of the world as a whole, as the integration of all the rich parts uncovered by separate disciplines. This emergence represents the uncovering of a whole new body of knowledge about Nature—Nature in its inter-connectence—Nature as dialectic. And this emergence is parallel with the disappearance of all bourgeois relations which mask a strait-jacket, an organization based on the conscious integration of the conscious specializations of labour. State forms and class forms, superimposed on the immediate life of society, wither away, and with them wither the fatal gulf between theory and practice which they generated. The crisis of physics is solved by the emergence of a new world-view, as the condition of the shattering of the old.

But this organization of society, due to the division of labour, as the result of the emergence to consciousness of what was hitherto unconscious, is the result of the emergence of the class in whose womb the object has developed unconsciously the whole division of labour. When this class extends to include all society by sucking into it the separate consciousness—i.e. by becoming a conscious organization—then the new organization has become conscious, society has become classless, bour-geois relations have been abolished, and the real world-view as an integrated whole has appeared. This is the emergence of the proletariat to power; then extension as a result of socialism to include all society. The crisis in physics is a part of the final crisis in bourgeois economy which gives rise to revolution and the creation of a new economy.

INDEX

Printed in the United States
by Baker & Taylor Publisher Services